International Conference on Nanomaterials and Biomaterials (ICNB 2017)

IOP Conference Series: Materials Science and Engineering
Volume 350

Amsterdam, The Netherlands
11 – 13 December 2017

ISBN: 978-1-5108-6281-4
ISSN: 1757-8981

Printed from e-media with permission by:

Curran Associates, Inc.
57 Morehouse Lane
Red Hook, NY 12571

Some format issues inherent in the e-media version may also appear in this print version.

Printed by Curran Associates, Inc. (2018)

For permission requests, please contact the Institute of Physics
at the address below.

Institute of Physics
Dirac House, Temple Back
Bristol BS1 6BE UK

Phone: 44 1 17 929 7481
Fax: 44 1 17 920 0979

techtracking@iop.org

Additional copies of this publication are available from:

Curran Associates, Inc.
57 Morehouse Lane
Red Hook, NY 12571 USA
Phone: 845-758-0400
Fax: 845-758-2633
Email: curran@proceedings.com
Web: www.proceedings.com

Table of contents

Volume 350

International Conference on Nanomaterials and Biomaterials (ICNB 2017)

11–13 December 2017, Amsterdam, The Netherlands

Accepted papers received: 16 April 2018
Published online: 14 May 2018

Preface

Papers

Nanomaterials and Composites

Surface Engineering

Biomaterials and Biomedical Engineering

IOP Conf. Series: Materials Science and Engineering **350** (2018) 011001 doi:10.1088/1757-899X/350/1/011001

PREFACE

It is our great pleasure to welcome you to 2017 International Conference on Nanomaterials and Biomaterials (ICNB 2017) which is held in Amsterdam, Netherlands during December 11-13, 2017. ICNB 2017 is dedicated to issues related to nanomaterials and biomaterials.

The major goal and feature of the conference is to bring academic scientists, engineers, industry researchers together to exchange and share their experiences and research results, and discuss the practical challenges encountered and the solutions adopted. Distinguished professors are invited to deliver keynote speeches regarding latest information in their respective expertise areas. It will be a golden opportunity for the students, researchers and engineers to interact with the experts and specialists to get their advice or consultation on technical matters, sales and marketing strategies.

This proceeding presents a selection from papers submitted to the conference from universities, research institutes and industries. All of the papers were subjected to peer-review by conference committee members and international reviewers. The papers selected depended on their quality and their relevancy to the conference. The volume tends to present to the readers the recent advances in the field of Nanomaterials, Biomaterials, Material Chemistry, Materials Processing Engineering and Electronic Materials etc.

We would like to thank all the authors who have contributed to this volume and also to the organizing committee, reviewers, speakers, chairpersons, sponsors and all the conference participants for their support to ICNB 2017.

Prof. Kwang Leong Choy
University College London, UK
December 12, 2017

Organizing Committee

Conference Co-Chairs

Prof. Carlo Montemagno, University of Alberta, Canada

Prof. Kwang Leong Choy, University College London, UK

Prof. Detlef W. Bahnemann, Leibniz University of Hannover, Germany

Program Chairs

Prof. Pankaj Vadgama, Queen Mary University of London, UK

Prof. Xuanhua Li, Northwestern Polytechnical University, Xi'an, China

International Technical Committee

Prof. Rafael Audino Zambelli, Federal University of Ceará, Fortaleza, Brazil

Prof. Habil. Simona Cavalu, University of Oradea, Romania

Prof. Antonio Zuorro, University of Rome, Italy

Prof. Takao Hanawa, Tokyo Medical and Dental University, Japan

Prof. Kazuo Umemura, Tokyo University of Science, Japan

Prof. Li Shang, Northwestern Polytechnical University, China

Prof. Arief BUDIMAN, Singapore University of Technology & Design, Singapore

Prof. Indriana Kartini, Universitas Gadjah Mada, Indonesia

Prof. K. BYRAPPA, Mangalore University, India

Prof. O.L. Shanmugasundaram, K.S. Rangasamy College of Technology, India

Assoc. Prof. Kenneth Loh, University of California, San Diego, USA

Assoc. Prof. Wael Zaki, Khalifa University, Abu Dhabi, UAE

Assoc. Prof. Fei Xu, Northwestern Polytechnical University, China

Assoc. Prof. Keyu Xie, Northwestern Polytechnical University, China

Assoc. Prof. Hua LI, Nanyang Technological University, Singapore

Dr. Eric Freeman, University of Georgia, USA

Dr. Ya Wang, Stony Brook University, USA

Dr. Dan Sathiaraj G, Technische Universität Dresden, Germany

Dr. Kelly Velonia, University of Crete, Greece

Dr. Avinash BAJI, Singapore University of Technology & Design, Singapore

Dr. DING Lu, Institute of Materials Research & Engineering, Singapore

Dr. Pinar Camurlu, Akdeniz University, Turkey

IOP Conf. Series: Materials Science and Engineering **350** (2018) 011003 doi:10.1088/1757-899X/350/1/011003

Peer review statement

All papers published in this volume of *IOP Conference Series: Materials Science and Engineering* have been peer reviewed through processes administered by the proceedings Editors. Reviews were conducted by expert referees to the professional and scientific standards expected of a proceedings journal published by IOP Publishing.

IOP Conf. Series: Materials Science and Engineering **350** (2018) 012001 doi:10.1088/1757-899X/350/1/012001

Development of a Post-CMOS Compatible Nanoporous Thin Film layer Based on Al_2O_3

Ö Dogan[1], A Buschhausen[2], C Walk[1], W Mokwa[3] and H Vogt[4]

[1] Fraunhofer Institute for Microelectronic Circuits and Systems, 47057 Duisburg, Germany
[2] Department of Microelectronics, University of Applied Sciences, 40476 Düsseldorf, Germany
[3] Department of Materials in Electrical Engineering 1, RWTH Aachen University, 52074 Aachen, Germany
[4] Department of Electronic Components and Circuits, University of Duisburg-Essen, 47057 Duisburg, Germany
E-mail: oezgue.dogan@ims.fraunhofer.de

Abstract. Porous alumina is a popular material with numerous application fields. A post-CMOS compatible process chain for the fabrication of nanoporous surface based on Al_2O_3 by atomic layer deposition (ALD) is presented. By alternately applying small numbers of ALD cycles for Al_2O_3 and ZnO, a homogenous composite was accomplished, for which the principle of island growth of ALD materials at few deposition cycle numbers was utilised. By selective texture-etching of ZnO content via hydrofluoric acid (HF) in vaporous phase at 40 °C and 10.67 mbar, a porous surface of the etch resistant Al_2O_3 could be achieved. TOF-SIMS investigations verified the composition of ALD composite, whereas AFM and high resolution SEM images characterised the topographies of pre- and post-etched samples. Pores with opening diameters of up to 15 nm could be detected on the surface after vaporous HF treatment for 2 minutes. The amount of pores increased after an etching time of 5 minutes.

1. Introduction

Porous alumina provides numerous applications in several domains of nanotechnology due to its relatively high strength and an improved thermal and chemical stability [1, 2]. It has been established in membrane technology [3 - 5], where pore size and pores amount play a significant role by modulating membrane properties, as a support for heterogeneous catalysts [6, 7] or also as evaporation masks [8]. Especially in the field of medical engineering and biotechnologies porous alumina functions as biocompatible coating for implantable devices, membranes and, especially the porous surfaces, as suitable cell interfaces [9 - 11]. Porous alumina is also well established in the field of resistive or capacitive measurement of relative humidity [12], in which pore size [13] and degree of porosity of the transducing material influence sensor sensitivity significantly by enlarging the surface for interaction with more water molecules [14]. The most common methods for synthetisation of porous Al_2O_3 are annealing at temperatures above 1000 °C [15, 16], anodization in liquid electrolytes [17, 18] or processing by sol-gel method with an additional annealing step [19, 20]. These methods have been well established in the past decades and show sufficient results. Nevertheless, they are aggressive for sensitive substrate materials, may result in inhomogeneous film thickness on complex surfaces and, thus, are not suitable for CMOS post-processing.

In order to avoid the above mentioned disadvantages of synthetisation methods, lower processing temperatures and a more isotropic deposition of the alumina can be applied by using atomic layer deposition (ALD). ALD is a special type of chemical vapour deposition (CVD) processes and ensures a highly conformal and dense thin film coating, even on non-planar, three-dimensional substrate surfaces with high aspect ratios [21]. Additionally, the deposition of Al_2O_3 by an ALD process is possible at temperatures below 300°C [22]. Unlike CVD processes, ALD utilizes an alternating exposure of two precursors. Thus, the substrate surface can be saturated with respective chemicals in each deposition cycle, so that film growth is self-limiting. This enables well-controlled film growth, with a specific growth rate, and layer thicknesses.
Based on mentioned properties of ALD layers; production of porous ALD ceramics cannot be implemented without additional process steps. A known method for generating porosity in an ALD layer is the deposition of a hybrid material by alternately applying an organic and inorganic precursor in terms of a molecular layer deposition (MLD). By post-deposition annealing or wet-etching procedures, the organic part of the ALD/MLD material can be removed so that a porous, inorganic oxide backbone is left [23, 24]. Nevertheless, the reproducibility of ALD processes is more manageable than MLD processes [25]. This is why inorganic precursors are considered in this paper, only. ALD of ceramics are well tested and established processes. Especially the deposition of Al_2O_3 with trimethylaluminum and water ($AlME_3/H_2O$) can be considered as the most typical and studied ALD process.
Therefore, a composite of Al_2O_3 and ZnO by ALD has been developed which shall be textured by treatment with vaporous hydrofluoric acid (vHF). Regarding [26, 27], a selective etching of the ZnO grain boundaries is expected, whereas the etch-resistant Al_2O_3 content shall remain. Furthermore, etching could occur at weak chemical bonds between Al_2O_3 and ZnO grain boundaries. Both mechanisms would lead to texturing of the composite and, thus, to a nanoporous composite layer (see figure 1).

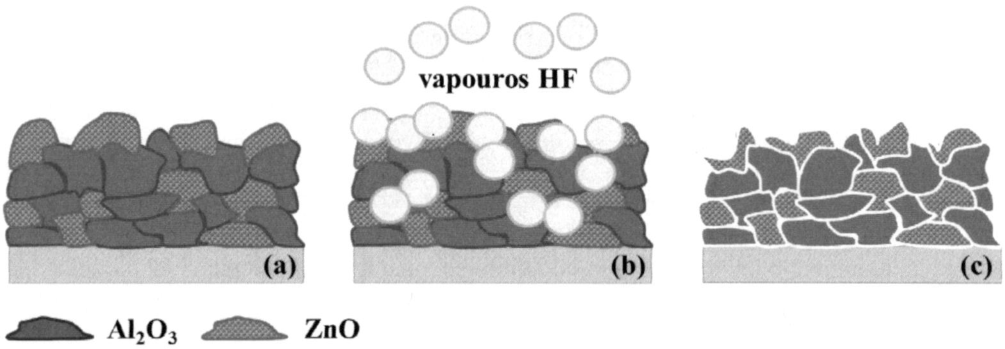

Figure 1. Schematic view of fabrication process of a nanoporous composite layer by ALD. (a) Al_2O_3-ZnO composite, (b) exposure to vaporous HF, (c) resulting nanoporous composite layer.

For a homogenous composite of both oxides, the principle of island growth at small numbers of deposition cycles has been utilised. If the growth of a metal oxide is substrate-inhibited, growth rate in the first deposition cycles can be much smaller than growth rates on ALD grown materials. Considering steric hindrances from precursor molecules and localised distribution of active sites on the substrate surface, island growth of respective materials can be achieved, if the amount of deposition cycles is small enough to prevent touching of neighbouring islands, which continuously grow with each additional deposition cycle [28].
In this paper, three ALD recipes have been investigated with TOF-SIMS to find a homogeneous Al_2O_3-ZnO composite. Most promising layers have been etched in vHF. AFM measurement and high-resolution SEM pictures compare pre- and post-etched results to verify the best route to fabricate a nanoporous surface.

2. Experimental details

In this work, three Al_2O_3-ZnO composite depositions were accomplished on silicon substrates with a thermal ALD equipment type "R200" from "Picosun Oy" which is a single-wafer reactor with vertical gas flow. Nitrogen (N_2) was used as carrier and purging gas. The deposition temperature was 270 °C. According to [28], Al_2O_3 islands grow starts at active defects within the substrate surface during the first ALD cycle. Islands grow laterally with each additional deposition cycles and start to touch each other after a certain number of cycles. Thus, island coalescence begins and results into a continuous film, if this certain cycle number is exceeded. Since this number is 16 in [28], smaller numbers of deposition cycles were chosen in this work. Used precursors were trimethylaluminum ($AlME_3$) and water (H_2O) for Al_2O_3 and diethylzinc (DEZ) and H_2O for ZnO. Every composite film was deposited using 0.1 s $AlME_3$ and 0.1 s H_2O pulses for Al_2O_3 with 6 s and 10 s purging steps, respectively. ZnO was deposited using 0.1 s DEZ and 0.2 s H_2O pulses with purging times of 6 s and 10 s. In order to define the upper limit of cycle numbers for a homogenous composition instead of a nanolaminate, different cycle numbers (x for alumina, y for zinc oxide) and oxide ratios were applied. One deposition period of Al_2O_3 first and ZnO second was defined as one super-cycle. Detailed process parameters are listed in table 1.

Table 1. Deposition cycle numbers, mixture ratio and super cycle numbers.

Sample #	x	y	Ratio x:y	Super cycle
1	8	8	1.00	35
2	12	10	1.20	25
3	15	12	1.25	21

Distribution of different oxide clusters within the composite was investigated by time-of-flight secondary ion mass spectrometry ("TOF.SIMS^5-100" from "ION-TOF" GmbH) which allows depth profiling with resolutions of < 1 nm per sputter step. An area of 10000 µm² was monitored and an average of intensity was calculated for detected fragments. Intensity trends with respect to sputter depth reveal information about the composition of Al_2O_3 and ZnO in the deposited ALD composites.

The isotropic and selective texture-etching of the composite contents via vHF is performed in the "Sacrificial Layer Release System" from "memsstar". This system comprises load lock, transfer module and process modules, of which one is capable of applying gaseous hydrofluoric acid (HF). In this process module, the pressure is controlled by the opening angle of the valve on the tube to the vacuum pump. For first experiments, the temperature, the pressure and total gas flows were hold constant at 40 °C. Higher pressures or lower process temperatures promote the condensation of water vapour. Thus, the catalytic effect of liquid water and total etching process is promoted as well [29]. Before the as described depositions were treated with HF in vapour phase (HF concentration 40 %), the texture-etching for ALD deposited ZnO was successfully tested at 10.67 mbar. On the contrary, Al_2O_3 did not exhibit any indications for successful etching at the same process pressure. ALD composite samples were treated with vHF for 2 and 5 minutes at 40 °C process temperature and 10.67 mbar process pressure.

Roughness of etched and not etched samples are compared by atomic force microscope (AFM) measurements ("D5000V" from "Veeco") recorded in taping mode. Scanning electron microscope (SEM) images of sample topographies were recorded with ultra-high resolution field emission SEM ("JSM-7500F" from "JEOL").

3. Results and discussion

3.1. TOF-SIMS

Composition of Al_2O_3 and ZnO contents is investigated by TOF-SIMS. Measured intensity signals of detected Al_2O_3 and ZnO fragments are presented in figure 2 – 5. As shown in figure 2, intensities of composite materials of sample 1 significantly decrease at the interface to the bulk material, where the intensity of detected bulk silicon increases at the same sputter depth. Figure 3 shows the enlarged depth within the range of ALD composite thickness (100 nm). Intensities of Al_2O_3 and ZnO are constant which results into a constant ratio, as well. This indicates that the amount of Al_2O_3 and ZnO clusters are constant at every sputter step through the whole oxide composite.

As shown in figure 4, the interface between ALD composite and bulk material of sample 3 is at about 166 nm. Detected intensities are obviously alternating during the whole sputter depth, which can be seen in the magnified range of composite thickness (figure 5). Based on period lengths of phase shifted intensities, a nanolaminate (or at least layers with significant predominant content of single materials) with Al_2O_3 layers of up to 5 nm thickness and ZnO layers of up to 3 nm thickness is extracted. SIMS results of sample 2 also showed phase-shifted intensities of Al_2O_3 and ZnO with smaller period lengths, which indicate thinner laminate-layers (approx. 2 nm). Since the intensity trends are very similar to the intensity trends of sample 3, they are not presented additionally.

Composites with 12 and 10 or 15 and 12 deposition cycles for Al_2O_3 and ZnO, respectively, showed nanolaminate properties. A vHF treatment would only lead to a texture-etching at the grain boundaries of the top ZnO film, whereas the underlying alumina film serves as an etch stop. Thus, a required porosity might not be achieved. Therefore, sample 2 and 3 are not further considered for etching steps and characterisations. Sample 1 showed continuously unchanged intensity signals over the whole ALD composite film which means that 8 deposition cycles of Al_2O_3 and ZnO, each, are suitable for achievement of island growth.

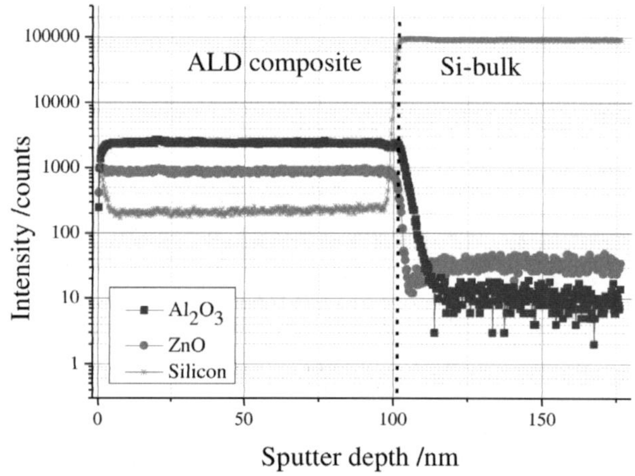

Figure 2. Sample 1 with each 8 deposition cycles, composite-bulk interface at 100 nm.

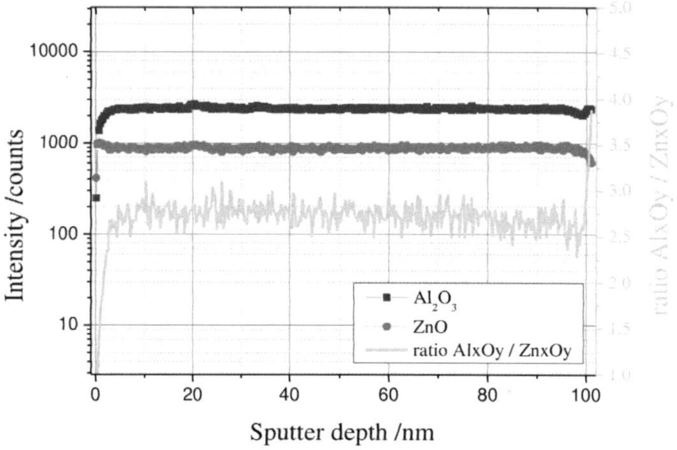

Figure 3. Sample 1 - magnified area with unchanged intensities of Al_2O_3 / ZnO and constant ratio thereof.

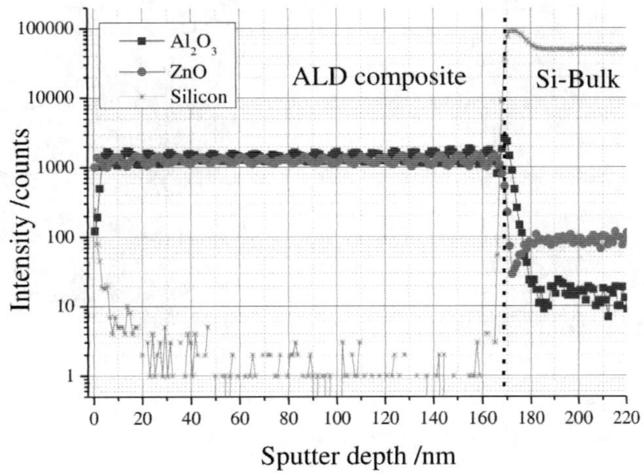

Figure 4. Sample 3 with 15 cycles Al_2O_3 and 12 cycles ZnO, composite-bulk interface at about 166 nm.

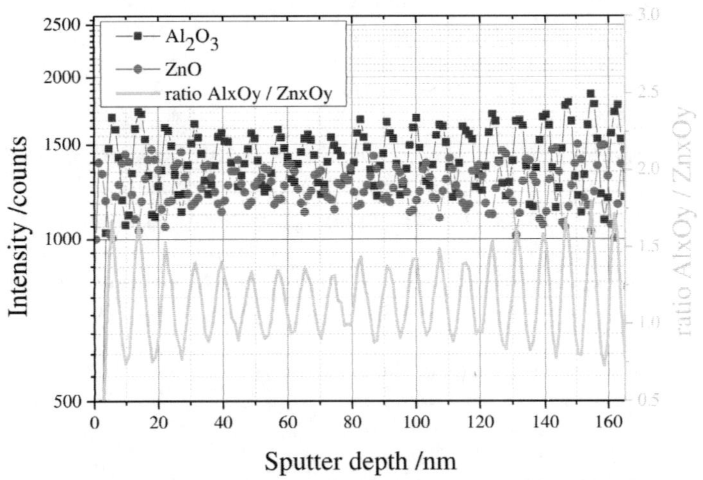

Figure 5. Sample 3 – magnified area with continuously phase-shifted intensities of Al_2O_3/ ZnO and alternating ratio.

3.2. AFM

AFM measurements were accomplished at tapping mode with a velocity of 0.3 μm per second and 256 measured lines (256 measurement points per line). Hill-like elevations of up to 6 nm were measured due to alternately applied ALD of two oxides and their different growth rate and behaviour. Figure 6 (a) presents the topography of the not etched ALD composite. Measured roughness was 1.15. This value is comparable with a roughness of 1.08 of the composite which was etched by vHF for 5 minutes (shown in figure 6 (b)). Even though roughness results are very similar, differences in granularity are obvious. The not etched sample showed a smaller granularity whereas the etched sample revealed bigger grain sizes. A reason for smaller grain sizes for the as-deposited composite is different growth behaviours of Al_2O_3 and ZnO. Both oxides have various growth rates as well on foreign substrate materials as on same ALD-grown materials. A mixture of both oxides with few deposition cycles leads to a composite with a granular topography due to continuous island growth while an increased number of deposition cycles allows the current material to grow more conformal.

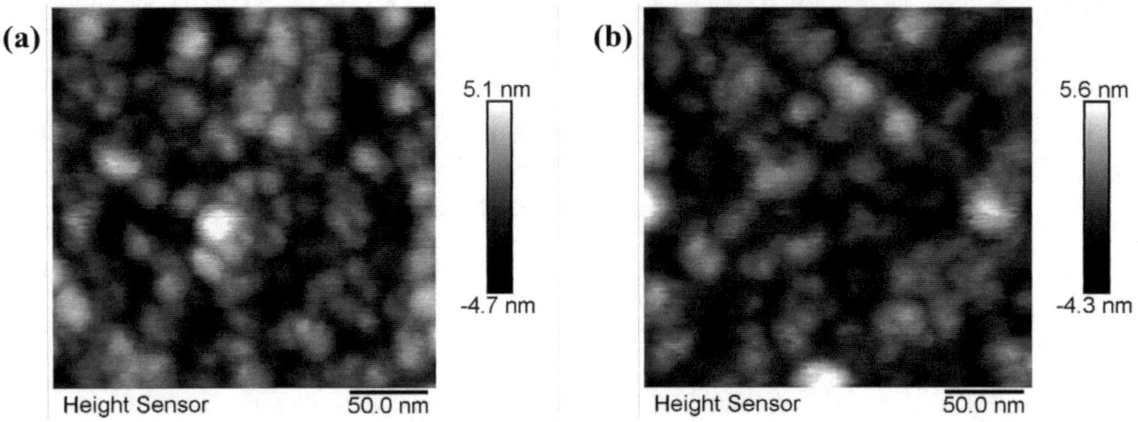

Figure 6. Topographies as measured by AFM of sample 1 (a) before etching and (b) after etching with vHF for 5 minutes at 10.67 mbar and 20 °C.

AFM images of sample 2 and 3 showed the same trend in granularity on the composite surface. The bigger the number of deposition cycles the bigger the hill-like elevations on the sample's surface. By applying vHF, the surface was texture-etched and the granularity seemed to increase. Porosity cannot be determined with the images of figure 6, since AFM cannot reveal information about a deepening, whose inner diameter is bigger than the opening diameter.

3.3. Ultra-High Resolution SEM

Uneven surfaces of not etched samples can be seen on high-resolution SEM images with a magnification factor of 200 k (figure 7). Granular surface of the untreated ALD composite can be seen on figure 7 (a), whereas (b) and (c) show vHF treated composites after 2 minutes and 5 minutes of etching, respectively. Already after 2 minutes (the shortest etching time used) pores could be seen. The number of pores seems to increase with increased etching time. Measured pore diameters varied from approximately 3 nm to 15 nm for 2 minutes and 5 minutes etched composites.

Figure 7. SEM images with magnification factor 200 k of (a) sample 1 (b) sample 1 after vHF treatment for 2 minutes and (c) sample 1 after vHF treatment for 5 minutes.

4. Summary and conclusion

Preliminary experiments verified a novel process for fabrication of porous ALD thin film layer based on alumina, which is deposited by a special ALD process, in this work. To reach this, a homogenous ALD composite with Al_2O_3 and ZnO was needed. By TOF-SIMS investigations of three different ALD recipes with various deposition cycle numbers, an upper limit of deposition cycle number could be defined for prevention of a nanolaminate. By applying 8 cycles of Al_2O_3 and 8 cycles of ZnO alternately, only island growth could be achieved, so that selective texture-etching of the ZnO content could lead to porosity on the surface of the ALD composite. Granularity and roughness were compared by AFM results. Even though roughness didn't change after vHF treatment, grain sizes on the etched surface seem to increase after texture-etching. High-resolution SEM images showed topographies of pre- and post-etched composites. Pores with diameters of up to 15 nm could be detected after a vHF treatment of 2 minutes. By applying a longer etching time, the amount of pores on the surface of the composite increased.

This work shows a process chain with only two fabrication steps for the development of a porous thin film layer which is based on alumina. Considering the application of humidity measurement, pores with diameters of approximately 2 - 10 nm are desired for detection of low relative humidity values since they promote trapping of gaseous water molecules (of the same size) more effectively [13]. Pores with diameters up to 15 nm could be processed by vHF treatment of an ALD composite with, alternately applied, 8 deposition cycles for each Al_2O_3 and ZnO. The amount of pores on the surface of the composite could be increased by increasing the etching time. As future work, the tailoring of pore size and degree of porosity of vHF treated ALD composites with smaller numbers of Al_2O_3 and ZnO deposition cycles and ratios, and various etching times will be investigated. It is assumed that a customised porous layer based on Al_2O_3 could be fabricated for humidity measurement, for example.

The presented process steps are compatible to CMOS by fulfilling temperature limits and choice of materials. The etching medium vHF can be applied as well if an appropriate etching barrier layer is used in order to protect the CMOS substrate [30]. Thus, this process chain is suitable for post-processing of all kind of CMOS chips.

5. References

[1] Poco J F, Satcher Jr. J H and Hrubesh L W 2001 Synthesis of high porosity, monolithic alumina aerogels *Journal of Non-Crystalline Solids* vol 285 pp 57-63

[2] Eklund P, Sridharan M, Singh G, Bottiger J 2009 Thermal Stability and Phase Transformations of γ-/Amorphous-Al_2O_3 Thin Films *Plasma Processes and Polymers* vol 6 pp 5907-5911

[3] Hsieh H P, Liu P K T and Dillman T R 1991 Microporous Ceramic Membranes *Polymer Journals* vol 23 (5) pp 407-415

[4] Sola L, Alvarez J, Cretich M, Swann M J, Chiari M, Hill D 2015 Characterization of porous alumina membranes for efficient, real-time, flow through biosensing *Journal of Membrane Science* vol 476 pp 128-135
[5] Bocchetta P, Chiavarotti G P, Di Q F, Gullo U, Sunseri C 2004 Process for manufacturing a porous alumina membrane for fuel cells, EP1391235 A2
[6] Patermarakis G and Pavlidou C 1994 Catalysis over Porous Anodic Alumina Catalysts *Journal of Catalysis* vol 147 pp 140-155
[7] Zapf R et al. 2003 Detailed characterization of various porous alumina-based catalyst coatings within microchannels and their testing for methanol steam reforming *Chemical Engineering Research and Design* vol 81 pp 721-729
[8] Masuda H and Satoh M 1996 Fabrication of Gold Nanodot Array Using Anodic Porous Alumina as an Evaporation Mask Japanese Journal of Applied Physics vol 35 (2) pp 126-129
[9] Gultepe E, Nagesha D, Sridhar S and Amiji M 2010 Nanoporous inorganic membranes or coatings for sustained drug delivery in implantable devices *Advanced Drug Delivery Reviews* vol 62 Issue 3 pp 305-315
[10] Brüggemann D 2013 Nanoporous Aluminum Oxide Membranes as Cell Interfaces *Journal of Nanomaterials* Article ID 460870
[11] La Flamme K E, Popat K C, Leonie L, Markiewicz E, La Tempa T J, Roman B B, Grimes C A, Desai T A 2007 Biocompatibility of nanoporous alumina membranes for immunoisolation *Biometarials* vol 28 pp 2638-2645
[12] Khanna V K and Naher R K 1986 Carrier-transfer mechanisms and Al_2O_3 sensors for low and high humidities *Journal of Physics D: Applied Physics* vol 19 pp 141-145
[13] Connolly E J, French P J, Pham H T M and Sarro P M 2002 Relative Humidity Sensors Based on Porous Polysilicon and Porous Silicon Carbide *IEEE Sensors* vol 1 pp 499-502
[14] Chou K-S, Lee T-K and Liu F-J 1999 Sensing mechanism of a porous ceramic as humidity sensor *Sensors and Actuators B* vol 56 pp 106-111
[15] Wang H A and Kröger F A 1989 Pore formation during oxidative annealing of Al_20_3- : Fe and slowing of grain growth by precipitates and pores *Journal of Material Science* vol 15 Issue 8 pp 1978–1986
[16] Furlan K P, Pasquarelli R M, Krekeler T, Ritter M, Zierold R, Nielsch K, Schneider G A and Janssen R 2017 Highly porous α-Al2O3 ceramics obtained by sintering atomic layer deposited inverse opals *Ceramics International* vol 43 pp 11260-11264
[17] Smith A 1974 Process for producing an anodic aluminum oxide membrane, US 3850762 A
[18] Chen Z, Jin M-C, Zhen C and Chen G-H 1991 Properties of Modified Anodic-Spark-Deposited Alumina Porous Ceramic Films as Humidity Sensors *Journal of the American Ceramic Society* vol 74 (6) pp 1325-1330
[19] Bagwell R B and Messing G L 1996 Critical Factors in the Production of Sol-Gel Derived Porous Alumina *Key Engineering Materials* vol 115 pp 45-64
[20] Farahmandjou M and Golabiyan N 2015 New pore structure of nano-alumina (Al_2O_3) prepared by sol gel method *Journal of Ceramic Processing Research* vol 16 (2) pp 1-4
[21] Ritala M and Niinistö J 2009 *Chemical Vapour Deposition: Precursors, Processes and Applications* vol 1, ed A C Jones and M L Hitchman pp 158-206
[22] Puurunen R L 2005 Surface chemistry of atomic layer deposition: A case study for the trimethylaluminum/water process *Journal of Applied Physics* vol 97 Article ID 121301
[23] Sundberg P and Karppinen M 2014 Organic and inorganic-organic thin film structures by molecular layer deposition: A review *Beilstein Journal of Nanotechnology* vol 5 pp 1104-1136
[24] Lausund K B and Nilsen O 2016 All-gas-phase synthesis of UiO-66 through modulated atomic layer deposition *Nature Communications* vol 7 Article ID 13578
[25] George S M, Yoon B and Dameron A A 2009 Surface Chemistry of Molecular Layer Deposition of Organic and Hybrid Organic-Inorganic Polymers *Accounts of Chemical Research* vol 42 (4) pp 498-508

[26] Hüpkes J, Zhu H, Owen J I, Jost G and Bunte E 2012 Instabilities in reactive sputtering of ZnO:Al and reliable texture-etching solution for light trapping in silicon thin film solar cells *Thin Solid Films* vol 520 pp 1913-1917

[27] Lee Y-J, Park H, Ju M, Kim Y, Park J, Ai D V, Hussain S Q, Lee Y, Ahn S and Yi J 2014 Improvement of haze ratio of DC (direct-current)-sputtered ZnO:Al thin films through HF (hydrofluoric acid) vapor texturing *Energy* vol 66 pp 20-24

[28] Puurunen R L and Vandervorst W 2004 Island growth as a growth mode in atomic layer deposition: A phenomenological model *Journal of Applied Physics* vol 96 pp 7686-7695

[29] Drysdale D, O'hara T and Wang C H 2011 Characterisation and comparison of water and alcohol as catalysts in vapour phase HF etching of silicon oxide films *Symposium on Design, Test, Integration Packaging of MEMS/MOEMS (DTIP)* pp 35-40

[30] Walk C, Chen Y, Vidovic N, Kuhl A, Görtz M and Vogt H 2016 Development of a low temperature SiC protection layer for post-CMOS MEMS fabrication utilizing vapour release technologies *Micro-Nano-Integration (6. GMM-Workshop)* pp 126-131

IOP Conf. Series: Materials Science and Engineering **350** (2018) 012002 doi:10.1088/1757-899X/350/1/012002

Phase Composition, Crystallite Size and Physical Properties of B_2O_3-added Forsterite Nano-ceramics

S Pratapa[1], A Chairunnisa[1], U Nurbaiti[1,2], W D Handoko[1]

1 Department of Physics, Faculty of Mathematics and Sciences, Institute of Technology Sepuluh November (ITS), Jl. Arief Rahman Hakim, Surabaya 60111, Indonesia
2 Department of Physics, Faculty of Mathematics and Natural Sciences, Semarang State University, Jl. Raya Sekaran Gunung Pati, Semarang 50221, Indonesia

Abstract. This study was aimed to know the effect of B_2O_3 addition on the phase composition, crystallite size and dielectric properties of forsterite (Mg_2SiO_4) nano-ceramics. It utilized a purified silica sand from Tanah Laut, South Kalimantan as the source of (amorphous) silica and a magnesium oxide (MgO) powder. They were thoroughly mixed and milled prior to calcination. The addition of 1, 2, 3, and 4 wt% B_2O_3 to the calcined powder was done before uniaxial pressing and then sintering at 950 °C for 4 h. The phase composition and forsterite crystallite size, the microstructure and the dielectric constant of the sintered samples were characterized using X-ray diffractometer (XRD), Scanning Electron Microscope (SEM) and Vector Network Analyzer (VNA), respectively. Results showed that all samples contained forsterite, periclase (MgO) and proto enstatite ($MgSiO_3$) with different weight fractions and forsterite crystallite size. In general, the weight fraction and crystallite size of forsterite increased with increasing B_2O_3 addition. The weight fraction and crystallite size of forsterite in the 4%-added sample reached 99% wt and 164 nm. Furthermore, the SEM images showed that the average grain size became slightly larger and the ceramics also became slightly denser as more B_2O_3 was added. The results are in accordance with density measurements using the Archimedes method which showed that the 4% ceramic exhibited 1.845 g/cm3 apparent density, while the 1% ceramic 1.681 g/cm^3. We also found that the higher the density, the higher the average dielectric constant, i.e. it was 4.6 for the 1%-added sample and 6.4 for the 4%-added sample.

1. Introduction

Forsterite (Mg_2SiO_4) can be considered as a functional material since it has a wide range potential of applications. Their interesting physical properties such as high melting point (approximately 1850 °C), low thermal expansion (below 2×10^{-5} °C^{-1}) [1], good chemical stability and high-temperature thermal insulation [2] make suitable for refractory component. Moreover, pure forsterite is known to exhibit thermal expansion behavior which fits in the solid oxide fuel cell (SOFC) system as it is comparable to that of the SOFC components [3]. Forsterite also has low dielectric constants (ε' of 6.5 and ε'' at an order of 10^{-4}) which match the requirement for microwave communication [4,5].

Forsterite solid ceramics were produced by several methods, including microwave sintering [6] and solid-state reaction [3]. The raw materials can be commercial magnesia products and commercial silica products [6, 7] or silica from natural sources [3,8]. There have been syntheses of nano-forsterite powders [9-11] in the literature, none of nano-forsterite ceramics, particularly those which were made from silica sand as the silica source.

IOP Conf. Series: Materials Science and Engineering **350** (2018) 012002 doi:10.1088/1757-899X/350/1/012002

Boria (B_2O_3) has been used as a sintering agent to favor the densification of ceramics during sintering [7,12]. Since boria melts at 480 °C, sintering a boria-containing nano-forsterite compacted powder at high temperature is a strategy to obtain nano-forsterite ceramics. Provided the sintering temperature is below the reaction temperature between boria and the components of the ceramics, an improved densification can be expected without significant grain or crystallite growth of the major phase.

Here, we report the synthesis and characterization of boria-added forsterite nano-ceramics. The raw materials were commercial magnesia and Indonesia local sand-derived silica. The effects of boria addition on the phase composition, density-porosity, crystallite and grain sizes, and dielectric constant of the forsterite ceramics were examined.

2. Materials and methods

This research used a purified natural silica sand which was taken from Tanah Laut, South Kalimantan as the source of silica and MgO powder (Merck) which contained some magnesium hydroxide ($Mg(OH)_2$). The silica sand was purified through screening-washing, milling, magnetic separation, immersion in a 2M chloride acid finally drying at room temperature to obtain a white silica powder. This powder was further processed with reaction using NaOH to obtain $NaSiO_3$ and then titration with HCl to gain amorphous silica powder. The water in the MgO powder was removed by heating at 400 °C for 30 mins. The amorphous silica and the treated MgO powders at a molar ratio of 1:2 were then mixed with 3 wt% addition of poly(vinyl) alcohol through ball milling for 3 h. The mixture was calcined at 950 °C for 1 h to produce forsterite nanopowder. Manual grinding was applied to the powder to give finer forsterite powder, mixed with B_2O_3 powder at 1, 2, 3 and 4% by weight as a sintering agent. The sintering was done to the uniaxially-pressed powders at 950 °C for 4 h to produce B_2O_3-added forsterite nano-ceramics.

The phase analyses were applied to the laboratory x-ray diffraction (XRD) data which were collected from all ceramic samples using an Empyrean PANalytical XRD. The XRD used a Cu target and was operated at 40 kV and 30 mA with 2θ-range between 15 and 65°2 θ and 0.017 °2 θ step-size.The phase composition analysis on the XRD data was done using Rietica software [13].

The characterizations of the ceramics included diameter shrinkage, porosity and density, Vickers hardness, thermal expansion and dielectric constant. The porosity and density of the ceramics were determined using the Archimedes method following the Australian Standard No. 1774.5-2014. A transmission phase-shift method has been employed to an Advantest R3770 vector network analyzer connected to an A-INFOMW WG P/N 90WCAN rectangular waveguide to acquire the dielectric properties of the ceramics at a microwave frequency from 8 – 12 GHz. SEM micrographs were also collected from the fractured surface of the sintered ceramics using FEI Inspect-S50.

3. Results and discussion

Figure 1 presents the XRD pattern (CuKα radiation) of the calcined powder. Phase identification showed that the powder contained forsterite (Mg_2SiO_4), proto-enstatite ($MgSiO_3$), periclase (MgO) which referred to PDF number 34-189,3-523, 43-1022 respectively. Further analysis using Rietveld 'ZMV' method revealed that the corresponding weight fractions of these phases were 86.8, 9.9, and 3.2%. It means, the synthesis has produced forsterite as the dominating phase through presumably direct reaction of $2MgO+SiO_2 \rightarrow Mg_2SiO_4$. Meanwhile, proto-enstatite was yielded through $MgO+SiO_2 \rightarrow MgSiO_3$. These results indicate that the reaction between silica and magnesia (a) was incomplete since there was some magnesia excess and (b) occurred in a non-homogeneous mixture. The latter can be concluded from the fact that forsterite, proto-enstatite, and periclase are simultaneously identified in the powder. This powder synthesis result is in accordance with such a forsterite synthesis where quartz powder was used as the raw material [3]. Further analysis using MAUD software, after an instrument broadening correction, showed that this powder exhibited a crystallite size of approximately 100 nm. Therefore, a relatively high content, nanometric forsterite powder has been successfully produced using the method mentioned above.

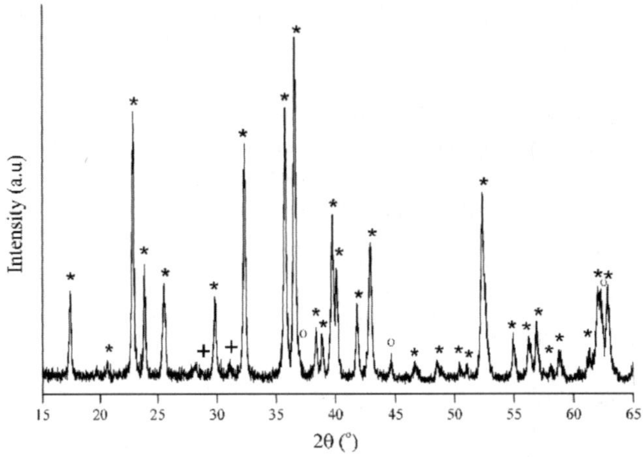

Figure 1. XRD pattern (CuKα radiation) of the calcined powder.
* = forsterite (Mg_2SiO_4); + = proto-enstatite ($MgSiO_3$);
o = periclase (MgO).

Then, XRD data analyses were done to provide the structure, the phase composition and the crystallite size information of forsterite in the sintered ceramics – the patterns are shown in Figure 2. In general, the ceramics contained only forsterite and periclase, except the 1%-added (A) sample which also had proto-enstatite. The figure shows the diminishing of proto-enstatite and reducing peak intensity of periclase, which, in general, suggests that the amount of added boria affects the formation of forsterite. Table 1 displays the weight fraction of forsterite in the B_2O_3-added nano-forsterite ceramics. It is obvious that addition of 4 wt% of boria results in a nearly pure forsterite ceramic, which is not achieved in the powder. It is argued, then, that boria has favored the formation of forsterite. Sintering at 950 °C for 4 h allowed the melted boria to homogenize the distribution of the silica (or more exactly the proto-enstatite) and magnesia particles to form forsterite through $MgO + MgSiO_3 \rightarrow Mg_2SiO_4$. The results also imply that besides the identified periclase crystallites there should be some amorphous silica excess in the ceramic samples. To summarize, the addition of boria caused the improvement of the forsterite formation in the ceramics.

Table 1 also presents the crystallite size of forsterite from further XRD data analyses of the ceramics. The addition of boria from 1% to 4% has increased the crystallite size from 104 nm to 164 nm. It appears that the boria addition cannot fully withstand the crystallite growth of forsterite. The presence of melted boria during sintering not only homogenized the distribution of particles from different phases but also promoted crystallite growth of the same phase. Nevertheless, the 4% sample still exhibited sub-nanometric characteristics.

The density-porosity characters of the ceramics are also presented in Table 1. An increase in density (therefore a decrease in porosity) with the addition of boria is evident. It is worthing to note that the ceramics are still very porous, presumably due to the low sintering temperature. Figure 3 shows the SEM micrographs which support this results. Reduction of porosity is apparent but not significant.

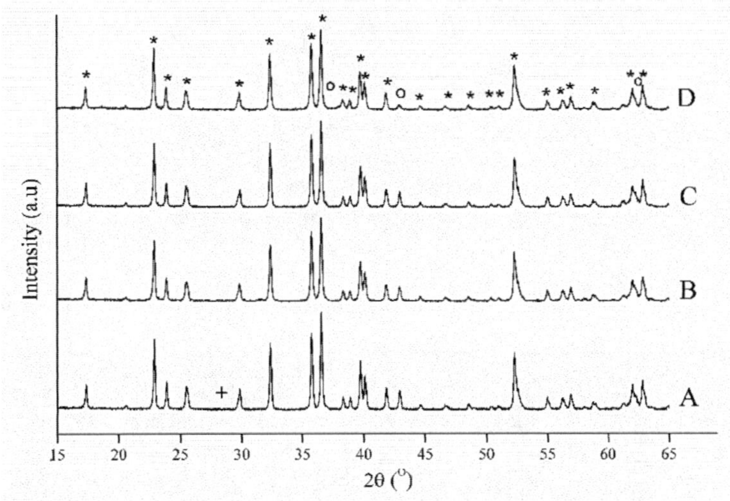

Figure 2. XRD pattern (CuKα radiation) of the B_2O_3-added ceramics.
* = forsterite (Mg_2SiO_4); + = proto-enstatite ($MgSiO_3$);
o = periclase (MgO).

Table 1. Weight fraction and crystallite size of forsterite in the B_2O_3-added nano-forsterite ceramics and their apparent density-porosity characteristics.

Sample	Weight Fraction (%)			Crystallite size (nm)	Density (g/cm^3)	Porosity (%)
	Forsterite	Periclase	Proto-enstatite			
A	95.3±0.7	4.5±0.3	0.2	104±13	1.68	44.3
B	95.6±0.8	4.4±0.3	0.0	110±6	1.70	38.2
C	95.7±0.3	4.3±0.3	0.0	118±19	1.82	36.9
D	99.1±0.8	0.9±0.1	0.0	164±24	1.85	25.5

Considering higher sintering temperature is possible, yet the crystallite size would be sacrificed as a result of a synthesis trade-off.

Figure 4 depicts the dielectric constant ε_r plots against frequency of the ceramics. Clearly, the dielectric constant of the ceramics is frequency-dependent. There are, for example, three distinct peaks for each sample at 8.4, 10.5 and 11.5 GHz. Moreover, the ranges of the dielectric constant value are 4.0-7.5, 4.0-7.5, 4.5-7.5 and 4.5-7.8 for sample A, B, C, and D respectively. The average dielectric constant was 4.6 for the 1%-added sample and 6.4 for the 4%-added sample. These ranges are slightly different and their fluctuations are insignificant. In addition, these values are low, as also found by others [3,14,15]. Therefore, the ceramics exhibited dielectric properties which make them appropriate for millimeter-wave communication applications.

4. Conclusions

This study showed that forsterite-based nano-ceramics can be synthesized via a solid reaction between amorphous silica, which was derived from natural silica sand, and magnesia powders. The nano-metric size of the forsterite can be retained with the largest size reached a value of 164 nm after 4 wt% boria addition. The addition also caused a significant increase in the weight fraction of forsterite and a decrease in porosity. Interestingly, although slightly frequency-dependent, the real dielectric constant of the ceramics are similar in their values.

IOP Conf. Series: Materials Science and Engineering **350** (2018) 012002 doi:10.1088/1757-899X/350/1/012002

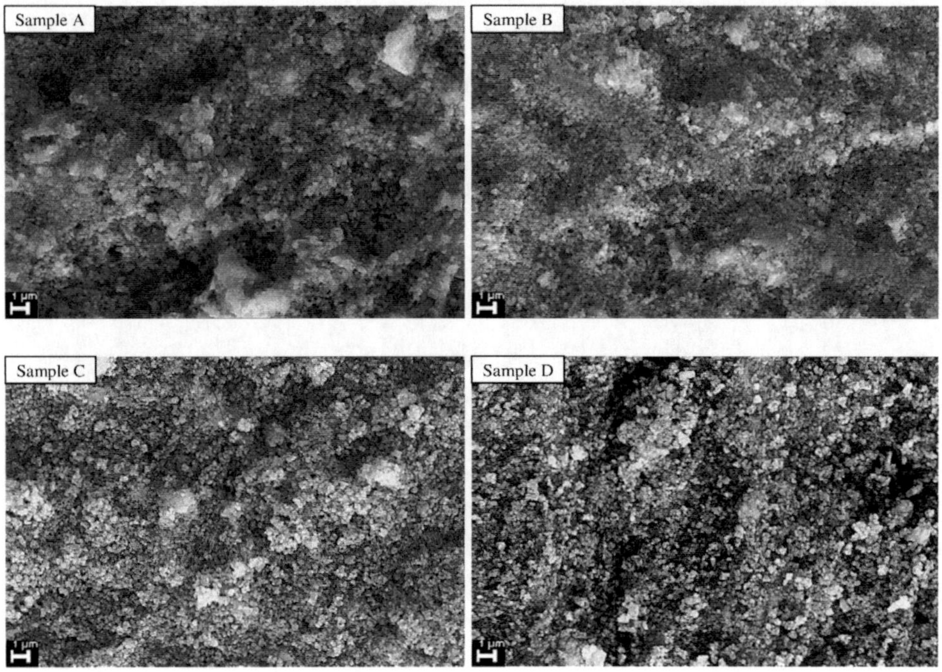

Figure 3. SEM micrographs of fractured surface of the B_2O_3-added ceramics.

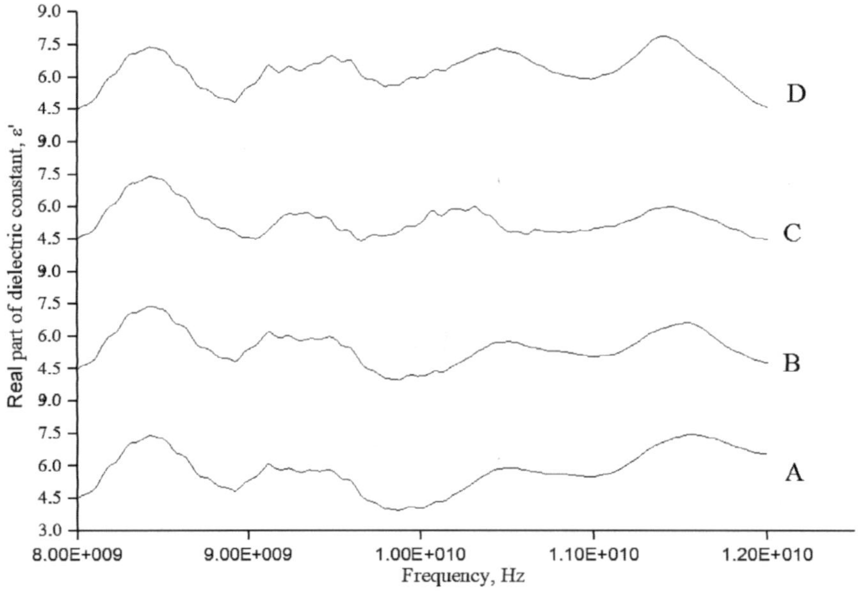

Figure 4. Frequency-dependent dielectric constant ε_r of the B_2O_3-added ceramics.

5. References

[1] Choudhury N, Chaplot S L and Rao K R 1989 Equation of state and melting point studies of forsterite *Phys Chem Minerals* **16** 599–605

[2] Delvaux P, Desrosiers L and Gouin M 1991 Forsterite and its use as insulating material *US Patent* No *US4985164 A*
[3] Pratapa S, Handoko W D, Nurbaiti U and Mashuri 2017 Synthesis and characterization of high-density B_2O_3-added forsterite ceramics *Ceramics International* **43** 7172–6
[4] Ohasto H, Tsunooka T, Ando M, Ohishi Y, Miyauchi Y and others 2003 Millimeter-wave dielectric ceramics of alumina and forsterite with high quality factor and low dielectric constant *Journal of the Korean Ceramic Society* **40** 350–3
[5] Tsunooka T, Ando M, Suzuki S, Yasufuku Y and Ohsato H 2013 Research & developments for millimeter-wave dielectric forsterite with low dielectric constant, high Q, and zero temperature coefficient of resonant frequency *Japanese Journal of Applied Physics* **52** 09KH02
[6] Barzegar Bafrooei H, Ebadzadeh T and Majidian H 2014 Microwave synthesis and sintering of forsterite nanopowder produced by high energy ball milling *Ceramics International* **40** 2869–76
[7] Fan G, Zhou H and Chen X 2016 Optimized sintering temperature and enhanced microwave dielectric performance of $Mg_2B_2O_5$ ceramic *J Mater Sci: Mater Electron* 1–5
[8] Reaney I M, Wise P, Ubic R, Breeze J, Alford N M, Iddles D, Cannell D and Price T 2001 On the temperature coefficient of resonant frequency in microwave dielectrics *Philosophical Magazine A* **81** 501–10
[9] Fathi M H and Kharaziha M 2008 Mechanically activated crystallization of phase pure nanocrystalline forsterite powders *Materials Letters* **62** 4306–9
[10] Kamran A H R, Moztarzadeh F and Mozafari M 2011 Synthesis and characterization of high-pure nanocrystalline forsterite and its potential for soft tissue applications *Advanced Composites Letters* **20** 41–7
[11] Saberi A, Alinejad B, Negahdari Z, Kazemi F and Almasi A 2007 A novel method to low temperature synthesis of nanocrystalline forsterite *Materials Research Bulletin* **42** 666–73
[12] Liu C-Y, Tsai B-G, Weng M-H and Huang S-J 2013 Influence of B_2O_3 additive on microwave dielectric properties of Li2ZnTi3O8 ceramics for LTCC applications *International Journal of Applied Ceramic Technology* **10** E49–56
[13] B.A. Hunter, Rietica, newsletter of international union of crystallography, Comm.Powder Diffr. 20 (1998) 21
[14] Bernardo E, Fiocco L, Giffin G A, Di Noto V and Colombo P 2014 Microstructure development and dielectric characterization of forsterite-based ceramics from silicone resins and oxide fillers *Advanced Engineering Materials* **16** 806–13
[15] Sano S, Saito N, Matsuda S-I, Ohashi N, Haneda H, Arita Y and Takemoto M 2006 Synthesis of high density and transparent forsterite ceramics using nano-sized precursors and their dielectric properties *Journal of the American Ceramic Society* **89** 568–74

Acknowledgement
We are indebted to the Ministry of Research Technology and Higher Education of the Republic of Indonesia and Institute for Research and Community Services ITS Surabaya for the support of research funding provided to SP through the PUPT-2016 No. 01751/IT2.11/PN.08/2016 and PBK-2017 No. 528/PKS/ITS/2017.

IOP Conf. Series: Materials Science and Engineering **350** (2018) 012003 doi:10.1088/1757-899X/350/1/012003

Numerical investigation of porous materials composites reinforced with natural fibers

M Chikhi[1], F Chekired[1] N Metidji[1] and F Mokhtari[1]

[1]Unité de Développement des Equipements Solaires, UDES/Centre de Développement des Energies Renouvelables, CDER, Bou Ismail, 42415, W. Tipaza, Algérie.
E-mail: chikhimourad06@yahoo.fr, chikhi.mourad@udes.dz

Abstract. The present article tends to predict the effective thermal properties of porous biocomposites materials. The composites matrix consists on porous materials namely gypsum and the reinforcement is a natural fiber as date palm fibers. The numerical study is done using Comsol software resolving the heat transfer equation. The results are fitted with theoretical model and experimental results. The results of this study indicate that the porosity has an effect on the Effective thermal conductivity biocompoites.

Nomenclature

k_f : thermal conductivity of fibers,
K_m : thermal conductivity of matrix
ETC : effective thermal conductivity $(W.m^{-1}.K^{-1})$
DPF_3 : date palm fibers of 3mm length
DPF_6 : date palm fibers of 6mm length
V_c: composite volume
V_{ce} : elementary cell volume
x, y, z : elementary cell dimensions
X, Y, Z : biocomposite dimensions
r : fiber radius, L : fiber length

1. Introduction

Reducing the energy consumption in building is a current major problem which interests a several researchers [1, 2]. Among solutions to resolve this problem is the using material which deals with thermal insulation [3-5]. The use of biomaterials containing natural fibers as a reinforcement becoming a necessity considering their numerous advantages such as the combination of good mechanical, thermal, and acoustic properties those allow these types of materials to be used for different applications [6, 7].

Generally, the introduction of NF induces a porous material. The recourse to the porous material and composites is justified by the fact that they present the advantage of lower heat conduction and a high strength. The parameters such as pore size, porosity and pore distribution, particle size and particle distribution make the heat conduction in the porous material and composites very complicated to predict [8].

The effective thermal conductivity (ETC) is very important to define the insulation materials degree because thermal expansion by disordered temperature distribution and temperature gradient causes stress concentration around pore and composite. The conventional models for heat transfer in porous material and composite are not perfectly adequate because they are simplified [9-11].

The theoretical modeling of (ETC) of the composite is a very difficult task and requires knowledge of numerous parameters which are sometimes difficult to obtain (parameters related to microstructure, thermal contact resistance, etc.) [12, 13]. Other work on mathematical modeling, experimental characterization on a microscopic scale, and Microstructural analysis can improve the model predicting of the properties of composite materials. Therefore, it can be noted that, no theoretical model is complete to predict the thermal conductivity of Composites with precision, which makes numerical simulation indispensable for validating theoretical models.

Several models have been proposed for modeling and predicting the effective thermal conductivity of porous materials. Yue *et al* [14] used the face centered cubic model (FCC) to predict the ETC of the polymer matrix reinforced by metallic fillers. Chikhi *et al* [9] studied also the reinforcement of polymer materials with metallic fillers in aim to increase the ETC of composites. The results show that at high concentration of fillers, the numerical model diverges to the experimental and theoretical model results. This trend may be due to the resistance contact in the composites. Asakuma [8] investigated the factors that determine the effective thermal conductivity of porous structures and composites, by the homogenization method. Calmidi et al. [15] presented a one-dimensional thermal conduction model by considering the porous structure as a two-dimensional. Djoudi *et al.* [16] investigated the ETC of DPF reinforced gypsum materials, using auto-coherent homogenization method. They concluded that this method gives a good results for small fraction of fibers (>2%*w*).

The experimental used techniques and the modeling approach are based on the fact that the adobe is clearly a bio-composite material [17]. The composite material in which the matrix is the wet soil and the filler is a set of tubular straws. The simulations of thermal effective property were performed using COMSOL™ software. The main aim of this paper is to investigate numerically the ETC of biocomposite material based on gypsum material reinforced with DPF. The simulations are performed using Comsol Multyphysics software in stationary stat. The numerical results of this work are validated with theoretical models of predicting ETC and experimental results obtained previously [3].

2. Materials process

The table 1 illustrates the thermophysical properties of different materials used in this study. The thermophysical properties of the biocomposite based on gypsum reinforced on date palm fibers (DPF) were determined experimentally using the Ct-meter Apparatus. The detail of the experimentation process is given in previous work [3].

Table 1. Thermal properties of materials used in the simulation

Thermophysical properties	k (W.m^{-1}.K^{-1})	ρ (Kg.m^{-3})	Cp (J/kg.K)
Gypsum	0.44	1130	1400
DPF	0.085	388	930
Air	0.025	1.225	0.34

2.1. Theoretical models to predict thermal conductivity

Regarding the literature several models have been proposed to predict the thermal conductivity of fiber-filled composites in aim to fit experimental and numerical data with various models [18]. Theoretical models used in this study are: Parallel and series models, Maxwell, Hatta and Taya, Hashin and Shtrikman models [9].

2.2. Descriptive of modeling with COMSOL software

In order to model the transfer within the composite, we consider an elementary cell corresponding to a fiber considered a cylindrical shape (l, r) centered in a parallelepiped (PL) matrix Figure 1. The temperature field in the composite material is defined by solving numerically the Laplace equation using a finite element formulation with the following boundary conditions. The two faces perpendicular to the direction of the heat flow are isothermal at temperatures of T_1 and T_2 respectively. The faces parallel to the direction of the heat flow are adiabatic. The transfers by radiation and convection are negligible; the thermal contact resistance between matrices and charges is negligible; the dispersion of the fibers in the matrices is homogeneous;

2.2.1. Modeling without porosity account

The Figure 1 presents the elementary cell with limit conditions.

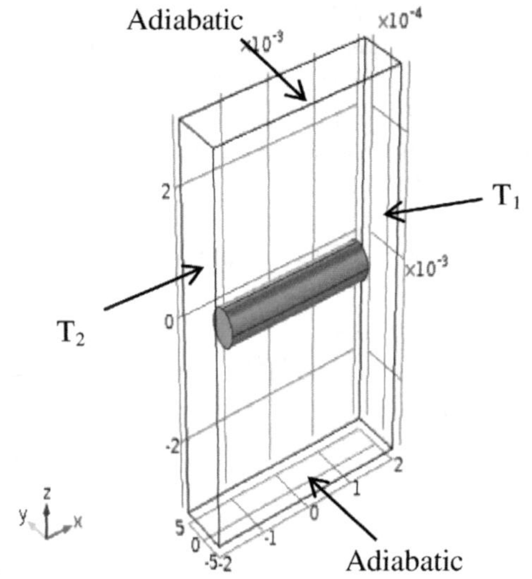

Figure 1. Elementary cell with limit

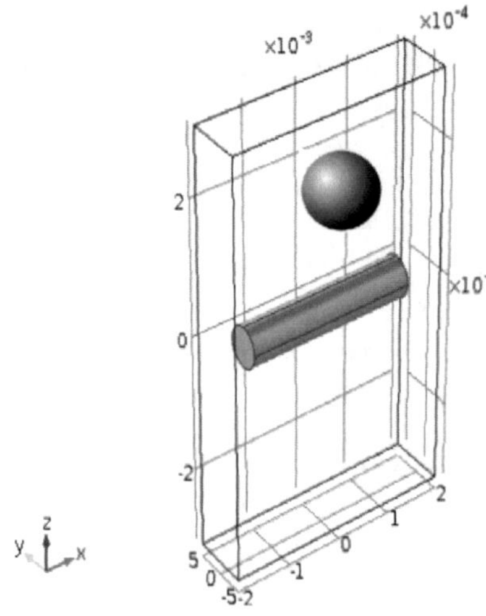

Figure 2. Elementary cell with pores

The heat flux that passes through the hot face towards the cold face is calculated from the integration of the Laplace equation (equation 1) applied to the boundaries of the elementary cell of the composite on the y direction.

$$Q = \int_A k \frac{\partial T}{\partial y} dxdz \quad (1)$$

With: Q is the heat flux, and k the effective ther mal conductivity of the biocomposite. x et z present the surface exchange.

To simplify the problem and minimize computation time, some assumptions are used:

$$V_c = X * Y * Z \quad (2)$$

$$V_{ce} = x * y * z \quad (3)$$

$$V_f = L * 2\pi r^2 \quad (4)$$

$$V_{ce} = V_{comp}/N \quad (5)$$

With:

L: the fiber length; r: the fiber radius and N the fibers number in the biocomposite ;

Taken into a count that the fiber is included in the gypsum (x>L; y>r; z>r).

$$k_{eff} = \frac{Q}{A}\frac{L_y}{(T_1 - T_2)} \quad (6)$$

Where the surface exchange A= x × z (m²); y is calculated, $\Delta T = T_1 - T_2$ (°K) is the temperature variation. The effective thermal conductivity is calculated using the Eq. 7.

$$k_{eff} = (Q * y)/(x * z * (T_1 - T_2)) \quad (7)$$

2.2.2. Modeling with porosity account

For the case of the simulation taken into account the void, the presence of the void in the biocomposites is mentioned as a sphere of radius R. The thermal properties of the void are similar to the air properties mentioned below. The Figure 2 presents the elementary cell with void.

3. Results and discussions

The main objective of this work is to investigate the void effect on the thermal conductivity of the composites

3.1. Modelization results without porosity account

The Figure 3 Presents the results comparison between numerical and theoretical models with the experimental results of the effective thermal conductivity of biocomposites reinforced with fibers of 3mm and 6mm.

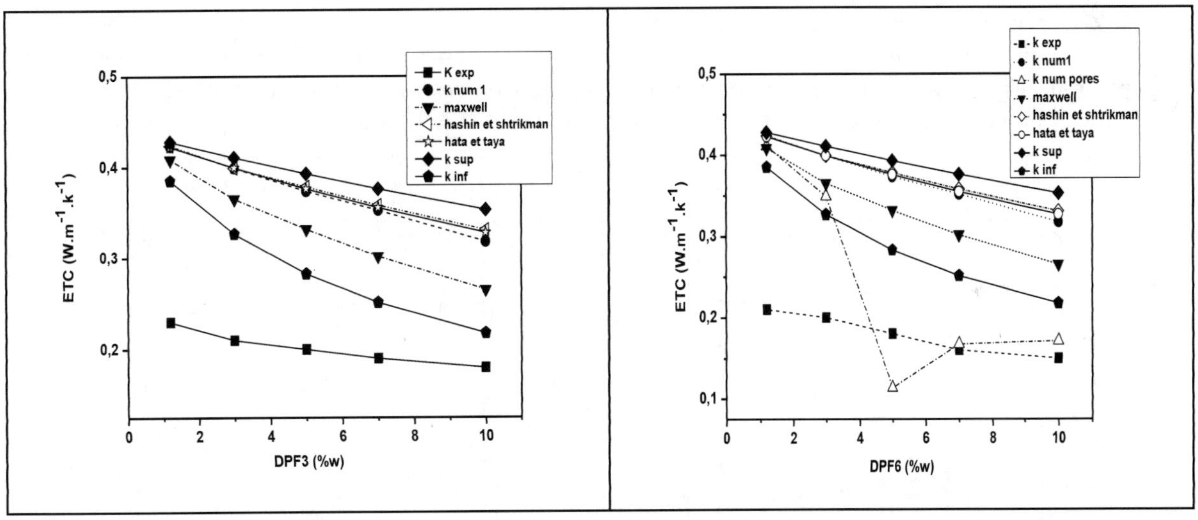

Figure 3. Comparison of thermal conductivities results of biocomposites; a) DPF_3; b) DPF_6

It can be seen from Figure 3, that the increase of the fiber mass concentration induces a non-linear decrease of the effective thermal conductivity of biocomposites reinforced with DPF length of 3 and 6 mm. This decrease is mainly due to the relatively low thermal conductivity of the inclusions (date palm fibers) compared to the matrix (gypsum) one. A similar behavior has been reported in the work of Djoudi et al. [16] who studied numerically the effect of the addition of date palm fibers on the thermal properties of gypsum concrete using the self-coherent homogenization method. The results showed that the effective thermal conductivity of the composite decreases as a function of the fraction of the fillers (date palm fibers). Several authors report this result using natural fibers as reinforcement of composites [1, 16]. An opposite trend is reported by Chikhi et al [9] who shows that the *k* of polymer based composite increases increasing the % of the fillers, which indicates that the effective thermal conductivity depends on the thermal conductivity of the inclusions [5, 9].

Comparing the numerical results with the theoretical models prediction of k_{eff}, it is observed that for the numerical values of the effective thermal conductivity are included between the lower and upper limits. It is noted that the Maxwell model does not predict the correctly the numerical results of effective thermal conductivity. This result may be due to the fact that the Maxwell model is based only on the thermal conductivities of the constituents, charge and matrix. While, the models Hashin and Shtrikman and Hata and Taya present a good estimation. This is related to the fact that these models introduce parameters related to the inclusions geometry. It should be noted that the Hashin and Shtrikman model agree with numerical results for fiber mass concentrations up to 10%. A low error of 4% at 10%*w* of fibers is recorded for HG/DPF_3 and 3% for HG/DPF_6. This result shows that the model of Hashin and Shtrikman predicts the k_{num} without voids. Moreover, the model of Hata and Taya is the model that gives the best estimate of the effective thermal conductivity of our bio-composite compared to the other predictions given by the other theoretical models. It should be noted that this model is based not only on the thermal conductivity of the inclusions but also on parameters related to the geometry of the fibers as in our case, short fibers oriented randomly. For comparison with experimental results, the both numerical and theoretical models results diverge. This result demonstrates that an important parameter is not taken into account in numerical and theoretical models. Since the matrix and fibers used in this study are porous materials, it is imperative to introduce the void parameter in the numerical calculations. As known the porosity effect strongly the thermal properties of biocomposites [17]. The next part of this study will take into account the porosity parameter.

3.2. Results of modeling with porosity account

In this section, the numerical study taken into account the presence of voids in bio-composite based on gypsum reinforced with DPF. Figure 4 presents a comparison between the numerical and experimental results as a function of the weight fraction of 3mm and 6mm fibers length

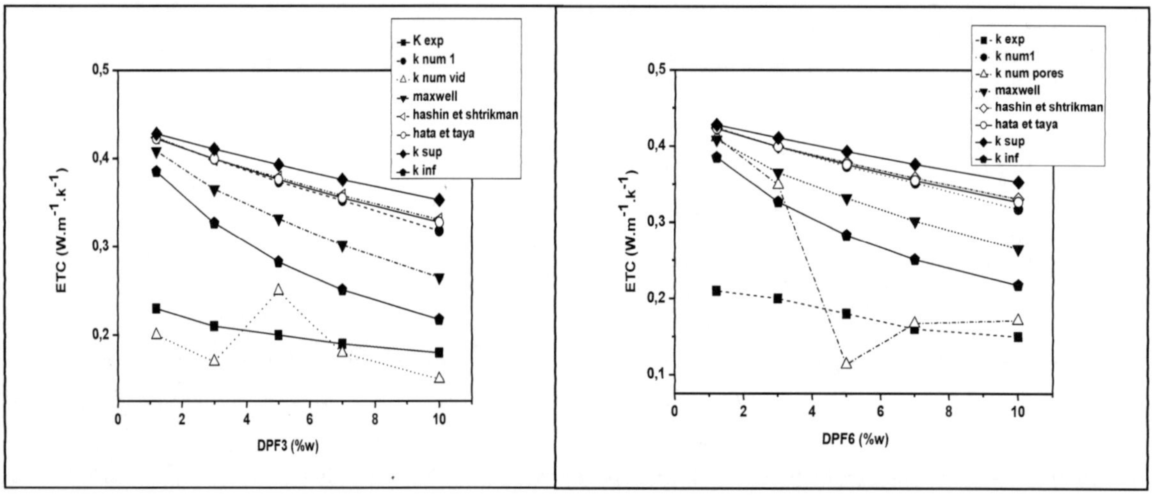

Figure 4. Effective thermal conductivity with voids; a) DPF 3mm, b) DPF 6 mm.

It is observed from Figure 4a, that the numerical results are close to the experimental values with an estimated error of 14% except for the concentration of 5%*wt* of DPF with a considerable value. This result may be the consequence of fibers distribution in gypsum.

In addition to the void parameter, the calculated error is mainly affect firstly, by the contact resistances between fiber/fiber and fiber/matrix. Secondly, because in reality when the fiber fraction increases the interaction between fibers increases, hence the formation of a chain of thermal conduction of the fibers which results in an increase in the thermal conductivity of the bio-composite. These two phenomena are not taken into account in numerical modeling which assumes a homogeneous distribution of the fibers in the matrix [9, 14].

Figure 4b indicates that the numerical results of the thermal conductivity in the second diverge from the values obtained by the theoretical models or beyond the 3% concentration of the fibers by 6 mm. This behavior is justified by the fact that theoretical models do not take into account the porosity parameter existing in the constitutions of bio-composites. This result is in line with the result works of [19] which indicates that the porosity effect the ETC of composite material.

4. Conclusion

In this study, the ETC of the gypsum matrix loaded with date palm fibers was studied numerically. The finite element method using Comsol Multypysics software was established to estimate the ETC of bio-composites studied in this work. On the one hand, the numerical results are in line with theoretical models results and diverge to the experimental ones for the case that the porosity does not taken into account. On other hand, the simulation taken into account pores of biocomposites, presents a good correlation with experimental results and diverge from the theoretical ones. This trend is due to the fact that the theoretical models do not considering several parameters which affect the ETC, as porosity, orientation morphology of fillers. The fibers size has an important effect on the ETC of biocomposites.

Reference:

[1] Korjenic A, Zach J, Hroudova J 2011 *Energy and Buildings* **43** 2518.
[2] Hajj N E, Dheilly R M, Aboura Z, Benzeggagh M, Queneudec 2011 *Industrial Crops and Products* **34** 921.
[3] Chikhi M, Agoudjil B, Boudenne A and Gherabli A 2013 *Energy Build* **66** 267.
[4] Chikhi M 2016 *Energy and Buildings* **129** 589.
[5] Benmansou N, Agoudjil B, Gherabli A, Kareche A, Boudenne A 2014 *Energy Build* **81** 98.
[6] Ali M, Li X, Chouw N 2013 *Mater Design* **44** 596.
[7] Yang HS, Kim DJ, Kim HJ 2003 *Bioresource Technol* **86** 117.
[8] Yusuke A and Yamamoto T 2013 *Computer Assisted Methods in Engineering and Science* **20** 89.
[9] Chikhi M, Agoudjil B, Boudenne A and Gherabli A 2013 *Journal of Thermoplastic Composite Materials* **26** (3) 336.
[10] Griesinger A, Spindler K, Hahne E 1999 *Int. J. Heat and Mass Trans* **42** 4363.
[11] Hasselman H and Johnson L F 1987 J. *Composite Materials* **21** 508.
[12] Bruggeman D A 1935 *Ann Phy* **24** 636,.
[13] Cheng S C and Vachon R I 1969 *Int. J. Heat. Mass. Transfer.* **12** 249.
[14] Yue C, Zhang Y, Hu Z, Liu J and Cheng Z 2017 *Microsystem Technolologies* **16** 633.
[15] Calmidi V and Mahajan R L 1999 *J Heat Transfer ASME* **121** (2) 466.
[16] Djoudi A, Khenfer M, Bali A and Bouziani 2014 *Journal of Adhesion Science and Technology* DOI: 10.1080/01694243.2014.948363
[17] Gines A A, Karkri M, Lefebvre G, Horn M, Solis J L, Monica M G 2011 *j.cscm* doi:10.1016/j.cscm.2017.02.001.
[18] Wattanakul K, Manuspiya H, and Yanumet N 2011 *J. Compos. Mater.* **45**(19) 1667 .
[19] Cernuschi F, Ahmaniemi S, Vuoristo P and Mantyla T 2004 *J. Eur. Cera. Soc.* 24 p 2657–2667.

IOP Conf. Series: Materials Science and Engineering **350** (2018) 012004 doi:10.1088/1757-899X/350/1/012004

Effect of Long Selenization Time on Co Films under a Low Temperature of 300 °C to Synthesize a Nanostructure $CoSe_2$ and Optical Properties

PO-FENG WU[1,a], JEN-BIN SHI[2,b*], BO-CI CHENG[2,c], CHENG-HAN WU[2,d], HSUAN-WEI LEE[2,e], HSIEN-SHENG LIN[2,f], FU-CHOU CHENG[3,g], KUAN-PING CHEN[2,h]

1College of general education, Hung-Kuang University, No. 1018 Sec. 6, Taiwan Boulevard, Shalu District, Taichung City 4330, Taiwan

2Engineering, Feng Chia University, No.100, Wenhwa Rd., Seatwen, Taichung 40724, Taiwan

3Professor & Director in Stem Cell Center, Department of Medical Research, Taichung Veterans General Hospital, No. 160, 3rd Section, Taichung Harbor Road, Taichung 40705, Taiwan

Email:[a]gavinwu0325@gmail.com,[b*]jbshi@fcu.edu.tw,[c]bochi0020@gmail.com,[d]sasabi96541 @gmail.com,[e]rocklee3001g@gmail.com, [f]shemglin@ms63.hinet.net, [g]fucheng@vght.gov.tw, [h]debby200598@gmail.com

Abstract. This work investigates a simple and non-toxic method to transform pre-deposited amorphous Co film into $CoSe_2$ films at a fixed, low temperature of 300 °C. Single $CoSe_2$-phase films having good crystallinity were obtained at a selenisation time ≧ 24 hours. A nanostructure $CoSe_2$ having two different nano-morphological layers was observed. The $CoSe_2$ films (72 hours) observed a large absorption and a direct band gap.

1. Introduction

Transition metal chalcogenides/dichalcogenides MX_2 (M = Mn, Fe, Ni, Co; X = S, Se, Te) has been investigated on their optical, magnetic, and electrical properties [1–4]. Many theoretical and experimental reports have been done on the pyrites MX_2 [5–7]. Due to the application in, rechargeable battery electrodes [8] and their potential applications in such areas as spintronic devices pyrites MX2[9]. CoSe2 and CoSe have received increasing appeal as potential cathodic catalysts for fuel cells [10-13]. Fuel cells can directly reduce our dependence on oil, energy use, and harmful emissions, which are used for stationary, portable power, and transportation. In the other hand, the oxygen reduction reaction (ORR), used in an acid medium Pt-based material, is the best and most used catalyst.

However, Pt materials are too expensive, while Fe and Co metals are much cheaper and serve equally as potential catalytic centers towards ORR in the creation of solar cell material. The cathodic oxygen reduction in chalcogenides of various transition metals has been further reported [14-16].

As an important pyrite-type compound, $CoSe_2$ has been known as a potential cathodic catalyst [14-16]. However, the extraordinary optical capability of $CoSe_2$ has not much been investigated for its potential application as a solar cell material of $CuInSe_2$ [17,18].

In this study, we demonstrated a simple and non-toxic method for direct low temperature synthesis of the $CoSe_2$ films with nanostructures. Transition metal dichalcogenides have been composited by using the direct stoichiometric combination of component in evacuated quartz ampoule [7]. However, as a result of the limit of slow diffusion, the whole response needs to operate at a high temperature from 500 to 1200 °C, and intermittent grinding [7]. No report is available for $CoSe_2$ thin film growth by selenization of cobalt films. This process runs at a low temperature of 300 °C, but requires a great deal of time. In this study, our aim is to further investigate the synthesis, structure, composition, and morphology of $CoSe_2$ films. The evolution of the optical properties of $CoSe_2$ is also discussed.

2. Detail Experimental

The $CoSe_2$ thin films were deposited onto the glass by selenization. In our work, the DC magnetron sputtering system was used to fabricate Co precursors. Co thin films were grown on the glass substrates and the chamber was initially evacuated to a base pressure of about 2.0×10^{-5} Torr. The Ar working pressure was 2.0×10^{-3} Torr and the DC power was 70 W for 1 hour. The Co target (99.95 % purity) was then presputtered for 10 minutes to remove any contaminants, especially oxide layering, on the surface of the Co target. 600 nm Co films were formed on the substrate at a deposition rate of 10 nm/minute.

All samples were put into a glass tube with pure Se powder when the Co precursors were fabricated. The glass tube was evacuated by using a pump, then placed in a furnace. The samples were then selenized in the furnace at 300 °C for 2, 5, 10, 24, and 72 hours. The phase and crystallization structure were treated by X-ray diffraction (XRD, SHIMADZU XRD-6000) utilizing Cu K_α radiation. In the course of observing the structure and morphology, the samples were used by a field-emission scanning electron microscope/energy dispersive spectrometer (FE-SEM/EDS, HITACHI S-4800). The optical absorption analysis of the $CoSe_2$ thin film was measured on a UV/Visible/NIR spectrophotometer (HITACHI U-3501).

3. Results and Discussion

Figure 1 shows the XRD patterns of the Co films selenized as a function of selenization time. The Co films and pure Se were annealed in a vacuum ($< 2 \times 10^{-5}$ Torr) sealed tube at different selenization times (2, 5, 10, 24, and 72 hours). Usually, there are three stable phases and they are homogeneous at room temperature ($Co_{0.85}Se$, CoSe and $CoSe_2$) and two other possible compositions (Co_2Se_3 and Co_3Se_4) for cobalt selenides [19-22].

The XRD pattern (Figure 1) of Co films as the starting precursor (0 hour) displayed an amorphous structure that does not show any peaks in the XRD diffraction pattern. These XRD patterns are very monotonic, and below 10 hours of selenization, they also show the presence of only two phases, whose peaks are marked with closed stars (CoSe phase) and squares ($CoSe_2$ phase). Phase one was indexed to the CoSe phase, identified as JCPDS card number 15-464, which has a hexagonal structure.

The other diffraction peaks can be indexed to the phase of the $CoSe_2$ hexagonal structure, which matches with the standard data file of $CoSe_2$. The XRD patterns (selenization time of 10 hours) showed no Se peaks, and the CoSe and $CoSe_2$ (Co/Se ratio ~ 1.7) phases coexisted. As the selenization time increased, the thickness and crystallinity of the films also increased; their peaks were sharper, and their intensities were drastically enhanced. The XRD patterns (Figure 1) were completely transformed into $CoSe_2$ after further thermal treatment (selenization time $\geq$ 24 hours). XRD patterns also showed the appearance of single phase $CoSe_2$. The $CoSe_2$ peak intensities corresponded to those in PDF 89-2002, and showed a good crystallinity. No peaks from other phases could be detected.

Surface and cross-sectional morphological images of the precursor Co films and the Co-Se films prepared from different selenization times are shown in Figure 2. The composition of the films has been estimated from energy dispersive analysis of X-rays (EDAX). At selected areas, the Se/Co atomic molar ratios obtained from cross sectional EDAX analysis in the upper/lower layers and the thicknesses of films are shown in Figure 3, as a consequence of the different selenization times. The EDAX results, at the upper layers, revealed the following composition in atomic ratio, where Se/Co ratios are above 1: atomic ratio of Se/Co of ~ 1.3 for film selenized at 5 hours; and ~ 1.7 for film selenized at 10 hours. The films selenized at 5 or 10 hours have a deficiency of Se content (selenization time $\leq$ 10 hours). The reaction that leads to the formation of the mixture of CoSe and $CoSe_2$ is favored in the upper layers. There are many nano-dots in the surface layer at a selenization time of 2 hours (Figure 2c & 2d). A layered structure was observed (Figure 2e – 2h) for the selenization times of 5 and 10 hours. The upper layer is composed of a mixture of CoSe and $CoSe_2$ phases. The lower layer corresponds to the Co phase due to EDAX results (Se/Co ~ 0). This presumption may be acceptable because the selenization process starts at the surface. The film in the upper layers thickens with an increase in selenization time (Figure 3), due to more selenium atoms with high activity diffusing into the crystal lattice of the Co-Se system. The amorphous structure in the precursor films can completely evolve into granular $CoSe_2$ grains through long selenization time ($\geq$ 24 hours). The nano-sized scale of $CoSe_2$ grains can be observed clearly in the films. The average size of the grains is within a range of 10 to 60 nm. The SEM surface and cross-sectional views of selenized films at a selenization time of $\geq$ 24 hours (Figure 2j and 2l) also showed a layered nano-structure. The upper layer is composed of the $CoSe_2$ phase, which has a large uniform area and nano-porous crystallite structure ($CoSe_2$-p). The lower layer is composed of a dense, nano-sized cylindrical crystallite structure ($CoSe_2$-d), and closely corresponds to stoichiometric $CoSe_2$ phases. Conversely, the upper layer absolutely corresponds to non-stoichiometric $CoSe_2$ phases, but the Se was found to be richer with an Se/Co ratio of ~ 2.2. As shown in Figure 2i -2l, two layers can be clearly found on the nano-structure $CoSe_2$ film. Those results are consistent with the X-ray diffraction analysis of the sample with a phase corresponding to $CoSe_2$.

The optical characteristic of the $CoSe_2$ film for 72 hours was measurement by optical absorption utilizing UV spectrophotometer. The absorption coefficient (α) is connected to photon energy (*hv*) by the following equation [23]:

$$(\alpha hv)^{m} = A(hv - E_g) \qquad (1)$$

where h is the Plank constant, A is a constant, E_g is the band gap, and m takes the value 1/2 and 1/3 for indirect allowed and indirect forbidden, value 2 for direct allowed transition, and value 2/3 for direct

IOP Conf. Series: Materials Science and Engineering **350** (2018) 012004 doi:10.1088/1757-899X/350/1/012004

forbidden, respectively. The transition in the thin film can be determined by plotting $(\alpha hv)^m$ against photon energy hv for a suitable value of "m", which leads to having a direct line. The value of E_g is obtained by extrapolating the linear portion of the graph to intercept the photon energy axis.

The type of transition is neither indirect nor direct, owing to the fact that there are no straight lines in any of the plots for $(\alpha hv)^{2/3}$ versus hv, $(\alpha hv)^{1/3}$ versus hv, and $(\alpha hv)^{1/2}$ versus hv. After 72 hours of selenization, the $CoSe_2$ thin film is a direct allowed transition. Through extrapolation of the linear portion of the plots drawn for $(\alpha hv)^2$ versus hv to the energy axis, the band gap of the film was ascertained. The optical band gap is 1.51 eV, as depicted in Figure 4. A similar result for previous researchers in optical absorption measurement have reported by Maneeprakorn et al. [24], who have found that the direct band gap of orthorhombic $CoSe_2$ nanoparticles was 1.45 eV. Figure 4 (insert) displays the absorption coefficient values of the $CoSe_2$ film (for 72 hour samples), which are between 1.0×10^5 to 2.0×10^5 cm^{-1}

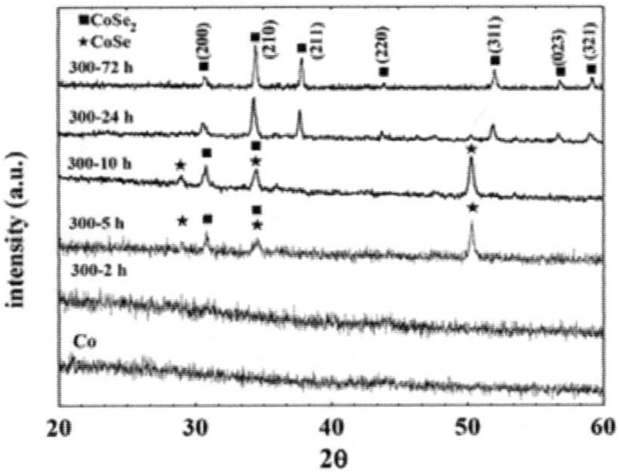

Figure. 1. The XRD patterns of samples selenized at 300 °C for different time.

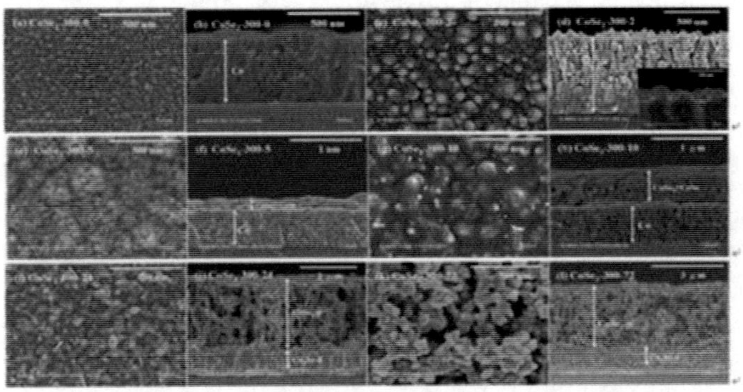

Figure. 2. FE-SEM images of top view and cross-sectional view of (a), (b) Cobalt thin films ($CoSe_2$-300-0); (c), (d) selenized at 300 °C for 2 hours ($CoSe_2$-300-2); (e), (f) selenized at 300 °C for 5 hours ($CoSe_2$-300-5); (g), (h) selenized at 300 °C for 10 hours ($CoSe_2$-300-10); (i), (j) selenized at 300 °C for 24 hours ($CoSe_2$-300-24); (k), (l) selenized at 300 °C for 72 hours ($CoSe_2$-300-72).

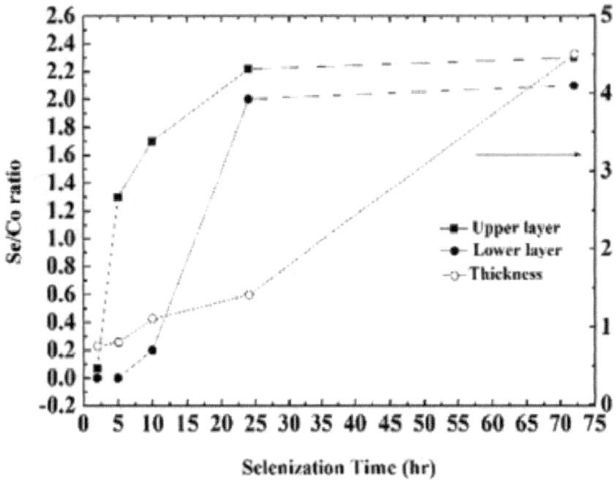

Figure. 3. The Se/Co atomic ratio of samples on the cross-section prepared at 300 °C for different length of time.

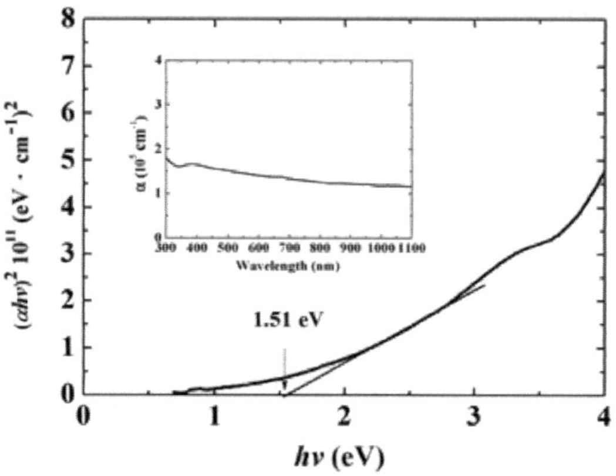

Figure. 4. The plots of (α hv)2 versus hv with absorption coefficient(inset) for CoSe2 thin films prepared at 300 ° C for 72 hours.

4. Conclusions

In summary, we demonstrate that this long lasting (up to ~ 3 days) method of selenization on Co films, using non-toxic pure Se at a low and fixed temperature of 300 °C, to obtain the nanostructed $CoSe_2$. When selenization occurs within a shorter time (≤ 10 hours), due to the limit of Se's very slow diffusion speed under low temperature, the $CoSe_2$ phase does not form completely. However, the Co film still exists on the bottom layers. Within this time frame, the selenization time rises and the Co-Se mixture films becomes thicker, and the CoSe and $CoSe_2$ phases become more dominant. With a selenization time ≥ 24 hours, a single $CoSe_2$ phase having good crystallinity was obtained throughout the films' thickening. SEM surface and cross sectional views show that $CoSe_2$ films have a nanostructure of two different morphological layers. Absorption measurements show that $CoSe_2$ films have a large absorption coefficient ($\alpha > 1.0 \times 10^5$ cm^{-1}) and a direct gap of about 1.51 eV. The nanostructure $CoSe_2$ film shows a direct band gap and high absorption coefficient.

5. References

[1] H.S. Jarrett, W.H. Cloud, R.J. Bouchard, S.R. Butler, C.G. Frederick and J.L. Gilson: Phys. Rev. Lett. Vol. 21 (1968), p. 617.
[2] P. Panissod, G. Krill, M. Lahrichi and M.F. Lapierre: Phys. Lett. A Vol. 59 (1976), p.221.
[3] S.M. Marcus and T.A. Bither: Phys. Lett. A Vol. 32 (1970) p. 363.
[4] K. Adachi, K. Sato and M. Takeda: J. Phys. Soc. Jpn. Vol. 26 (1969), p. 631.
[5] G. Krill and A. Amamou: J. Phys. Chem. Sol. Vol. 41 (1980), p. 531.
[6] V.H. McCann and J.B. Ward: J. Phys. Chem. Sol. Vol. 38 (1977), p. 991.
[7] H. Sato, F. Nagasaki, Y. Kani, S. Senba, Y. Ueda, A. Kimura and M. Taniguchi: Solid State Commun., Vol. 118 (2001), p. 563.
[8] L. Levy, N. Feltin, D. Ingert and M.P. Pileni: J. Phys. Chem. B Vol. 101 (1977), p. 9153.
[9] S.A. Wolf, D.D. Awschalom, R.A. Buhrman, J.M. Daughton, S.V. Molnar, M.L. Roukes, A.Y. Chtchelkanova and D.M. Treger: Science Vol. 294 (2001), p. 1488.
[10] R.A. Sidik and A.B. Anderson: J. Phys. Chem. B Vol. 110 (2006), p. 936.

[11] E. Vayner, R.A. Sidik, A.B. Anderson and B.N. Popov: J. Phys. Chem. C Vol. 111 (2007), p. 10508.
[12] D. Susac, A. Sode, L. Zhu, P.C. Wong, M. Teo, D. Bizzotto, K.A.R. Mitchell, R.R. Parsons and S.A. Campbell: J. Phys. Chem. B Vol. 110 (2006), p. 10762.
[13] M.R. Gao, Q. Gao, J. Jiang, C.H. Cui, W.T. Yao and S.H. Yu: Angew. Chem. Int. Ed. Vol. 50 (2011), p. 4905.
[14] L. Zhu, D. Susac, M. Teo, K.C. Wong, P.C. Wong, R.R. Parsons, D. Bizzotto, K.A.R. Mitchell and S.A. Campbell: J. Catal. 258, p. 235-242.
[15] Y.J. Feng, T. He and N.A.Vante: Chem. Mater. Vol. 20 (2008), p. 26.
[16] Y. J. Feng, T. He and N.A. Vante: Fuel Cells Vol. 10 No.1 (2010), p.77.
[17] J.B. Shi, Y.C. Chen, P.F. Wu, S.Y. Yang, T.K. Kang, P.C. Yang, C.L. Lin, S.L. Young and M.C. Kao: Res. Technol. Vol. 48 (2013), p. 94.
[18] J.B. Shi, Y.C. Chen, C.J. Chen and P.F. Wu: Cryst. Res. Technol. Vol. 47 (2012), p. 183.
[19] M. Hansen: Constitution of Binary Alloys, Geminuim Publishing Company, New York (1985), p. 502-503.
[20] X.H. Liu, N. Zhang, R. Yi, G.Z. Qiu, A.G. Yan, H.Y. Wu, D.P. Meng and M.T. Tang: Mater. Sci. Eng. B, Vol. 140 (2007), p. 38.
[21] J.H. Zhan, X.G. Yang, S.D. Li, Y. Xie, W.C. Yu and Y.T. Qian: J. Solid State Chem. Vol. 152 (2000), p. 537.
[22] J.F. Zhao, J.M. Song, C.C. Liu, B.H. Liu, H.L. Niu, C.J. Mao, S.Y. Zhang, Y.H. Shen and Z.P. Zhang: Cryst. Eng. Commun. Vol. 13 (2011), p. 5681.
[23] J. C. Osuma, C. I. Oriaku, F. I. Ezema and F. I. Ezema: Chalcogenide Lett. Vol. 6 (2009), p. 385.
[24] W. Maneeprakorn, M. A. Malik and P. O'Brien: J. Mater. Chem. Vol. 20 (2010), p. 2329.

Acknowledgments

This research was supported by the National Science Council of R.O.C. under grant No. MOST 105-2623-E-035-002-ET, MOST 105-2221-E-035-073, TCVGH-FCU1068202, MOST 106-2221-E-035-082 and the Precision Instrument Support Center at Feng Chia University.

IOP Conf. Series: Materials Science and Engineering **350** (2018) 012005 doi:10.1088/1757-899X/350/1/012005

Laser Cladding of Ti-6Al-4V Alloy with Ti-Al_2O_3 Coating for Biomedical Applications

A Mthisi[1,4], A P I Popoola[1], D I Adebiyi[2] and O M Popoola[3]

1 Department of Chemical, Metallurgical and Materials Engineering, Tshwane University of Technology, Pretoria West, South Africa.
2 Department of Metallurgical Engineering, Vaal University of Technology, Vanderbijlpark, South Africa.
3 Department of Electrical Engineering, Tshwane University of Technology, Pretoria West, South Africa.
4 Email: andriesmthisi@gmail.com

Abstract. The indispensable properties of Ti-6Al-4V alloy coupled with poor tribological properties and delayed bioactivity make it a subject of interest to explore in biomedical application. A quite number of numerous coatings have been employed on titanium alloys, with aim to overcome the poor properties exhibited by this alloy. In this work, the possibility of laser cladding different ad-mixed powders (Ti - 5 wt.% Al_2O_3 and Ti - 8wt.% Al_2O_3) on Ti-6Al-4V at various laser scan speed (0.6 and 0.8 m/min) were investigated. The microstructure, phase constituents and corrosion of the resultant coatings were characterized by scanning electron microscope (SEM), Optical microscope, X-Ray diffractometer (XRD) and potentiostat respectively. The electrochemical behaviour of the produced coatings was studied in a simulated body fluid (Hanks solution). The microstructural results show that a defect free coating is achieved at low scan speed and ad-mixed of Ti-5 wt. % Al_2O_3. Cladding of Ti - Al_2O_3 improved the corrosion resistance of Ti-6Al-4V alloy regardless of varying neither scan speed nor ad-mixed percentage. However, Ti-5 wt.% Al_2O_3 coating produced at low scan speed revealed the highest corrosion resistance among the coatings due to better quality coating layer. Henceforth, this coating may be suitable for biomedical applications.

1. Introduction

The human body biomechanical functions such as movement and body posture against gravity are made possible by musculoskeletal. Musculoskeletal is a biological system which is made of two sub-systems: the skeletal system and the muscular system [1]. However, in some circumstances tissue damage which could result from trauma or degenerative disease may be severe to a point that natural healing process would be inadequate [2]. With advances and innovations being made in the discipline of tissue engineering, it is possible to repair and restore the functions of a severely damaged bone tissue. In this regard, biomaterials known as scaffold would be required to bridge the gap and boost the osteogenic healing process [2]. A scaffold would serve a primary role of acting as an adhesion substrate for ingrowth cell, facilitates the delivery of drugs, provide temporary mechanical support to the newly grown tissue [3]. Hence, a material with unique characteristics such as biocompatibility, excellent corrosion resistance and low modulus is required to execute the aforementioned requirements expected from a scaffold. In this prospect, a material which satisfies almost all required properties and it is mostly used in biomedical applications is specifically titanium alloy (Ti-6Al-4V) [4-5]. Nevertheless, a scaffold made of Ti–6Al–4V is exposed to hostile corrosive body environment imposed by body fluid. This fluid emasculates the Ti–6Al–4V alloy capability to passivate and consequently leading dissolution of toxic ions like Al and V [6-7]. Moreover, the consequences of toxic ions lead to disruption of bioactivity behaviour between implant and tissues [8]. In case where articulation and movement occurs, and titanium alloy implant rubs against other materials such as bones; it would raise concerns. It has been reported that when titanium alloy slides against other hard or soft materials it suffers from wear [9]. Upon deterioration of protective oxide film due to wear, bare Ti6Al4V is subjected to rapid corrosion. This synergistic phenomenon is termed tribo-corrosion [10]. However, to overcome this setback and extend the life span of Ti–6Al–4V alloy incorporation of coatings with combination of excellent bioactivity, corrosion resistance, and biocompatibility is an effective method [11].

Fabrication of bioceramics like Ti and Al oxides (Al_2O_3) coatings may act as a barrier against diffusion of metallic ions from substrate into extracellular fluid [12]. However, because of their oxide nature, bioceramics are brittle and thus their application in load-bearing is restrained [13]. This challenge could be addressed by combining a metallic element with bioceramic and one metal which stands out to complement this combination is commercial pure titanium, as it exhibits characteristics such as corrosion resistance and good surface bioactivity [14]. Notably, the fabrication of Ti-Al_2O_3 composite on Ti-6Al-4V for biomedical application has been done by [12]. However, the authors used electrophoretic deposition technique. There are quite numerous techniques which could be employed to fabricate composites on titanium alloy. However, of all available surface treatments techniques, laser cladding has been proven to be a proficient method in fabricating (TiMMC) titanium metal matrix composites [15-16].

Zhang et al. [17] fabricated Al_2O_3/Ti-Al matrix intermetallic composite coatings on Ti–6Al–4V alloy using laser technique but emphasized more on mechanical properties of the coatings. Hence, this work is devoted to exploit the advantage of laser based coatings on the development of complex novel in - situ composite from admixed ratio of Ti-Al_2O_3 elemental powders that will mitigate the pressing issues related to Ti-6Al-4V alloy in biomedical applications. In addition, the influence of laser scan speeds on the electrochemical behaviour of Ti - 6Al - 4V in a simulated body fluid (Hanks solution) was investigated.

2. Experimental procedures

2.1. Materials

Ti-6Al-4V alloy of dimension (70 X 70 X 5 mm) acquired from VSMPOAVISMA Corporation were used as substrate in this work. The nominal chemical composition is as follows: 6.10 wt.% Al, 4.01 wt.% V, 0.15 wt.% Fe, 0.007 wt.% C, 0.12 wt.% O, 0.005 wt.% N, Ti, balance. Commercial pure titanium (99.9% purity) and alumina powders (99.9% purity) supplied by Sigma Aldrich, South Africa were chosen as coatings materials. These powders were mixed into the composition ratio of 95 wt.% Ti-5 wt.% Al_2O_3 and 92 wt.% Ti-8 wt.% Al_2O_3 using Turbula mixer at a mixing speed of 49 rpm for 6 h.

2.2. Laser deposition

Prior to laser cladding treatment, the surfaces of Ti6Al4V substrates were sandblasted and cleaned with acetone to remove contaminants from the surface and to enhance adsorption of the laser beam. Laser cladding was carried out using 4.4 kW continuous wave RofinSinar Nd:YAG equipped with an fiber optic beam delivery system. During laser deposition, composite powder was injected at a rate of 2 l/min into the melt pool formed by the laser beam during scanning of the substrate. Also, argon shielding gas was coaxially fed at 8 l/min. This was done in order to prevent oxidation as Ti has high affinity for oxygen. Moreover, to attain large clad surface area; multiple tracks were produced at 50% overlap. The laser spot diameter and power were kept constant at 2 mm and 900 W respectively, whereas, the scanning speed was varied between 0.6 and 0.8 m/min in order to distinguish the characteristics of clads obtained at various scan speeds.

2.3. Characterization of clad samples

2.3.1. Microstructure and phase analyses. After laser deposition treatment, the produced clad samples were prepared for microstructural analysis following standard metallographic procedure. The microstructures of the coatings were examined using SEM/EDS (JEOL JSM 7600 F FESEM), meanwhile the resulted phases were characterized using XRD (Philips P1710 Panalytical) with Cu target K_{α} radiation. The phases present were scanned and identified by X'Pert High score plus software.

2.4. Corrosion testing

Prior to corrosion testing, specimens were prepared by sectioning them into 1 cm^2 and cold mounted in an epoxy resin. Grinding and polishing using 1200-grit SiC papers and diamond suspension respectively were carried to achieve a well polished laser coating surface. Upon testing, specimen was immersed into a cell filled with Hank's solution (Sigma Aldrich) which resembles human body fluid. Electrochemical test was carried out using an Autolab potentiostat (PGSTAT30 computer controlled). An electrochemical cell setup was as follows: Hank's solution as electrolyte at 25 °C; three electrodes connected, the working electrode (sample), counter electrode (graphite rod) and silver/silver chloride electrode as a reference electrode (SCE). The corrosion potential (Ecorr) was determined by scanning at the rate of 0.01 V s^{-1} from a potential of −1.5 V to 1.5 V.

3. Results and discussion

3.1. Microstructural evolution

Figure 1 shows the SEM micrograph of Ti6Al4V substrate. It is observed that this alloy consists of two distinctive phase structures. The α-phase (darker phase) has an HCP structure whereas the β-phase (lighter phase) has a BCC structure. According to Adesina et al. [18], the α-phase is stabilized by Al whilst V stabilizes β-phase. Despite that these elements are crucial for stabilizing important phases, but their presence in this alloy poses human health risk if released into human system during service of this alloy in biomedical [16]. Henceforth, incorporation of composite such as Ti-Al_2O_3 coatings using laser may be valuable to suppress these toxic elements from being released.

Figure 2(a-d) shows the Optical micrographs of laser-clad coatings at varying scanning speeds of 0.6 and 0.8 m/min. Needle-like acicular structure known as α′ martensite is observed in all the coatings microstructure. The observed martensitic structure transformed from α + β due to high cooling rate as a result of laser processing [14, 19]. However, one could question why the martensitic structure distribution between coatings produced at various scan speed and ad-mixed powder was inconsistent. Possible explanations for this occurrence may be attributed to that addition of Al_2O_3 in Ti microstructures matrix may have prohibited the continuous nucleation of martensitic network. The coating materials possess thermal conductivities and melting temperatures as listed: titanium (17 W/mK) [20], 1604-1660 °C [21]; Al_2O_3 (6.3–28.7 W/mK), 2054 °C (2327 K) [22]. Hence, the differences in thermal properties perhaps led to inhomogeneous temperature gradient and cooling rates. Cracks and pores observed in Figure 2(c-d) affects the quality of the coating layer. Defects such as gas and pores forms at high scan speed because of gas bubbles entrapped during solidification [14].

3.2. Phase constituents

The XRD of coatings produced at varied laser scan speed and composite ad-mixed is displayed in Figure 3. As seen from this Figure, all the coatings reveal TiO_2 phase with high intensity. This phase formed probably because supplied heat by laser and quantity of O from air were adequate to prompt oxidation. Moreover, phases such as several types of aluminium oxides and titanium aluminide oxides were detected in all the coatings. The formation of these phases can also be witnessed in the other studies done by [12, 22, 23]. The penetration of O into the molten pool thus to form TiO_2 led to the establishment of the TiO_2-Al_2O_3 system. As a result of thermodynamically and kinetics mechanisms, TiO_2-Al_2O_3 system spontaneously favoured the formation of titanium aluminide oxides [23]. However, addition of more Al_2O_3 quantity possibly altered the formation of these phases regardless of varying laser scan speed. According to de Morais et al. [24] vanadium is highly toxic either in elemental and oxides state. Thus, the presence of V related phase in coatings (Ti-5 wt.% Al_2O_3 – 0.6 m/min and 0.8

m/min, Ti-8 wt.% Al_2O_3 – 0.8 m/min) would pose harmful effects if these coatings are considered for scaffold treatment. However, it could be noted that the absence of V related phases was achieved at the coating of Ti-5 wt. % Al_2O_3 deposited at 0.6 m/min scan speed. With this observation, the harmful effect of vanadium is thereby mitigated.

Figure 1. SEM micrographs of Ti-6Al-4V alloy.

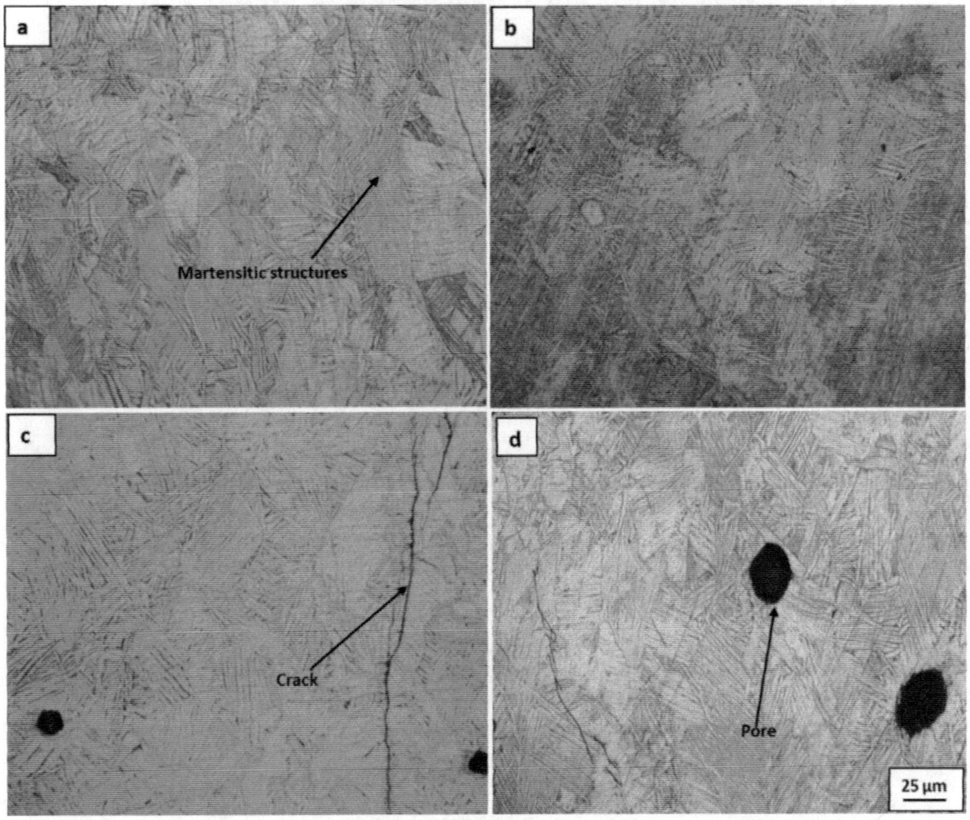

Figure 2. Optical micrographs of the coatings: (a) Ti-5 wt.% Al_2O_3– 0.6 m/min, (b) Ti-5 wt.% Al_2O_3– 0.8 m/min, (c) Ti-8 wt.% Al_2O_3– 0.6 m/min, Ti-8 wt.% Al_2O_3– 0.8 m/min.

3.3. Corrosion behaviour

Figure 4 shows the results of the electrochemical behaviour of the produced coatings and substrate. As it could be seen from Figure 4, that the potential and current density of the laser cladded coatings are superior compared with the Ti6Al4V alloy substrate. This implies that the fabricated coatings are vital in protecting the surface of the substrate from exposure to the electrolyte. As reported by Popoola et al. [16] the laser coating film on the surface acts as a cathode, which prohibits the movement of ions. However, the Ti-5 wt.% Al_2O_3 coating produced at 0.6 m/min exhibit the highest corrosion resistance among other coatings. According to Mokgalaka et al. [25] in certain environmental conditions, protective oxide film consisting of titanium dioxide (TiO_2) predominantly and also including aluminium oxide (Al_2O_3) is responsible for corrosion resistance. With relevance to this evidence, phases such as TiO_2 and Al_2O_3 coupled with titanium aluminide oxides phases observed in this work may be deemed responsible for achievement of superior corrosion resistance. Moreover, addition of more Al_2O_3 quantity (Ti-8 wt.% Al_2O_3) was supposed to yield better results. However, Al_2O_3 is porous thus this material creates a site for electrolyte accumulation and diffusion [12,26]. Thus, this is a reason addition of Ti-8 wt.% Al_2O_3 yielded inferior corrosion resistance than Ti-5 wt.% Al_2O_3. Also, Obadele et al. [14] observed that cracks and pores formed due to laser cladding at high scan speed such result in high corrosion density. This observation is also witnessed in this work.

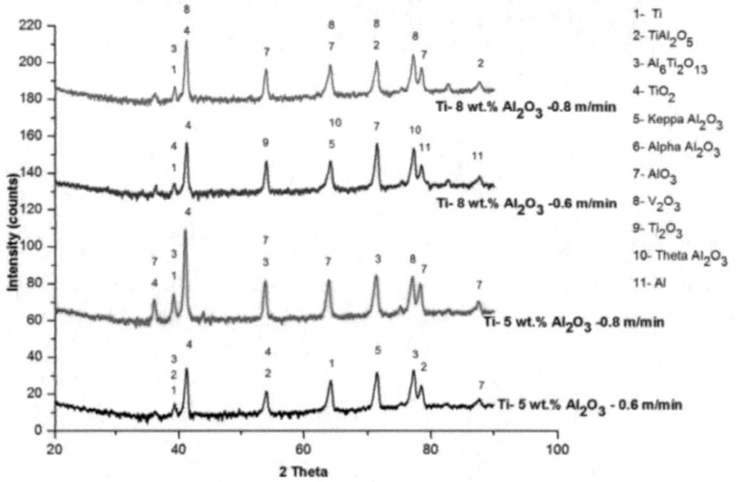

Figure 3. XRD of the coatings.

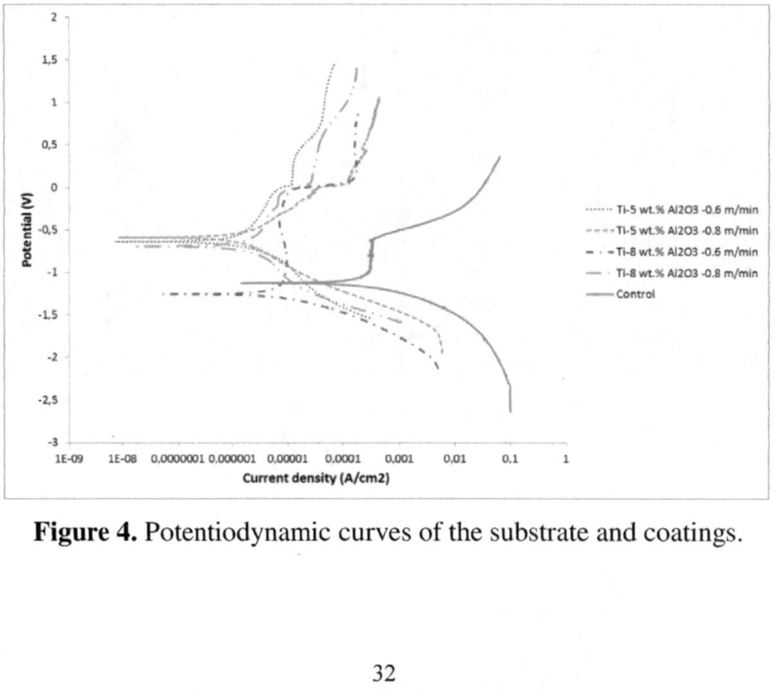

Figure 4. Potentiodynamic curves of the substrate and coatings.

4. Conclusion

Laser cladding coating of Ti-6Al-4V with Ti-Al_2O_3 is plausible. However, a defect free coating is achieved at coating of Ti-5 wt.% Al_2O_3 produced at laser scan speed of 0.6 m/min. The presence of martensitic structure coupled with TiO_2, Al_2O_3 and titanium aluminide oxides phases led to optimum corrosion resistance for the aforementioned coating. Henceforth, this coating could be considered for coating biomedical implants such as scaffold since it does not pose risk of vanadium dissolution into humans.

5. References

[1] Ren L, Qian Z and Ren L 2014 *J. Bionic. Eng.* **11** 159-175.
[2] Kanczler J M, Mirmalek-Sani S H, Hanley N A, Ivanov A L, Barry J J, Upton C, Shakesheff K M, Howdle S M, Antonov E N and Bagratashvili V N 2009 *Acta. Biomater.* **5** 2063-2071.
[3] Sarkar S K and Lee B T 2015 *Korean. J. Intern. Med.* **30** 279.
[4] Dinda G, Song L and Mazumder J 2008 *Metall. Mater. Trans. A* **39** 2914-2922.
[5] Alvarez K and Nakajima H 2009 *Materials* **2** 790-832.
[6] Pohrelyuk I, Fedirko V,Tkachuk O and Proskurnyak R 2013 *Corros. Sci.* **66** 392-398.
[7] Alves V, Reis R, Santos I, Souza D,Goncalves T d F, Pereira-da-Silva M, Rossi A, Da Silva L 2009 *Corros. Sci.* **51** 2473-2482.
[8] Ma J, Wang C Z, Ban C L, Chen C Z and Zhang H M 2016 *Vacuum* **125** 48-55.
[9] Allen C, Bloyce A and Bell T 1996 *Tribology International* **29** 527-534.
[10] Hacisalihoglu I, Samancioglu A, Yildiz F, Purcek G and Alsaran A 2015 *Wear* **332** 679-686.
[11] Sun T, Zheng N Y and Wang M 2012 *Int. J. Appl. Ceram. Technol.* **9** 517-528.
[12] Mahmoudi M, Maleki-Ghaleh H,Kavanlouei M and Aghaie E 2015 *Mater. Corros.***66** 479-485.
[13] Ibrahim M Z, Sarhan A A, Yusuf F and Hamdi M 2017 *J. Alloys Compd.***714** 636-667.
[14] Obadele B A, Olubambi P A, Andrews A, Pityana S and Mathew M T 2015 *J. Alloys Compd.***646** 753-759.
[15] Weng F, Yu H, Chen C and Dai J 2015 *Mater. Des.* **80** 174-181.
[16] Popoola A P I, Phume L, Pityana S and Aigbodion V S 2016 *Surf. Coat. Technol.* **285** 161-170.
[17] Zhang X W, Liu H X, Jiang Y H and Pham Thi H N 2013 *Journal of Inorganic Materials* **9** 023.
[18] Adesina O S, Mthisi A and Popoola A P I 2017 *Procedia Manufacturing* **7** 46-52.
[19] Balla V K, Soderlind J, Bose S and Bandyopadhyay A 2014 *J. Mech. Behav. Biomed. Mater.* **32** 335-344.
[20] Dong H 2010 *Tribological properties of titanium-based alloys* ed H Dong (Surface Engineering of Light Alloys) 58-80.
[21] Mahamood R M, Akinlabi E T, Shukla M and Pityana S 2013 *Mater. Des.* **50** 656-666.
[22] Chen Y, Wu D, Ma G, Lu W and Guo D 2013 Surf. Coat. Technol. **228** S452-S455.
[23] Niazi H, Yari S, Golestani-Fard F, Shahmiri M, Wang W, Alfantazi A and Bayati R 2015 *Appl. Surf. Sci.* **353** 1242-1252.
[24] de Morais L S, Serra G G, Palermo E F A, Andrade L R, Müller C A, Meyers M A and Elias C N 2009 *Am. J. Orthod. Dentofac. Orthop.* **135** 522-529.
[25] Mokgalaka M N, Popoola A P I and Pityana S L 2015 *Transactions of Nonferrous Metals Society of China* **25** 3315-3322.
[26] Maleki-Ghaleh H, Khalil-Allafi J, Khalili V, Shakeri M and Javidi M 2014 *Mater. Corros.* **65** 725-732.

Acknowledgements

This material is based upon the work financially supported by the National Research Foundation of South Africa. The National Laser Centre, CSIR, Pretoria, South Africa, is appreciated for laser facility. The authors also acknowledge the support from Tshwane University of Technology, South Africa.

IOP Conf. Series: Materials Science and Engineering **350** (2018) 012006 doi:10.1088/1757-899X/350/1/012006

Surface Modification of Polyester Fiber with Perfluorooctyltrimethoxysilane

Xiangcheng WANG [1, a], Yadong LIU‡[1, b], Dan LI‡[1] [1,c], Zihan TIE[1,d]

[1] Institute of Chemical Defense, Beijing 102205, China
E-mail:[a]lwjwhr@sina.com,[b]670315380@qq.com,[c]912317454@qq.com, [d] 303831338@qq.com

Abstract. An excellent modified polyester fiber was prepared via chemical grafting between polyester fiber and perfluorooctyltrimethoxysilane (FAS-17), or silane coupler (KH-570), or Titanate coupler (DN-101) in isopropyl alcohol aqueous solution. Volume ratio of isopropyl alcohol in aqueous solution was 50:50, the mass concentration of FAS-17 is 2%, reacting on polyester fiber modified for 24h at 60 ℃, the polyester fiber contact angle to water was 145 °, and the contact angle to peanut oil was 118°, with excellent performance of amphiphobic property. KH-570 and DN-101 modified polymer fiber to be hydrophobic properties nearly as FAS-17, but modified polyester fiber have no amphiphobic property.

1. Introduction

With the development of society and the progress of science and technology, filtering technology requirement is increasing, precision filtration and depth-type filtration and bulk leach put forward higher request to the filter and the filter materials [1]. The traditional filtration process is difficult to meet the requirements in the field of military and aerospace aviation oil, high grade lubricating oil production and oily waste water recovery technology. The modified polyester fiber has hydrophobic property, strong adsorption ability, and a property easy to back purge. And the fiber ball filter material made of polyester has large surface area because of its good fiber wire. Fiber ball filter material in the filter forms the filter layer, the filter layer hole becomes smaller and smaller along the direction of the flow, then the ideal big end up inter space distribution was formed. It was conformed that fiber ball filter layer is a good filter medium contribute to removing oil, it has a great advantage in treating sanitary sewage, polymer flooding in oil field and oil recovery, etc.

There are a lot of approach to modify fiber surface [2], such as alkali treatment, plasma treatment [3], photochemical treatment [4] and other methods, but some problems are still existed for example, vary

[1] ‡ Both authors contributed equally.

degrees strength loss, short validity period, high cost, technical difficulties, hard for mass production and other issues.

Chemical grafting modification on fiber surface is simple, effective and environmentally friendly. It is a hot topic in recent years. Polyester fiber can be dissolved in sodium hydroxide solution, caused by nucleophilic substitution reaction, the hydroxyl ion and carbonyl carbon of polyester fiber react to the anionic intermediate, then continue to fracture and generate hydroxyl and carboxyl end group [5].

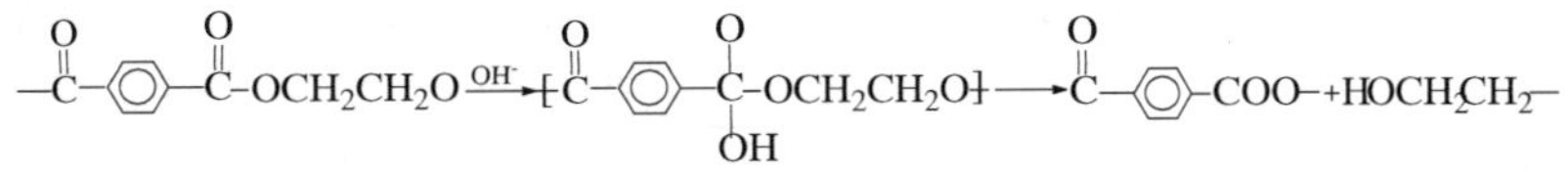

Grafting modification is hydrophilic/lipophilic monomer grafting polymerization on the polyester fiber surface initiated by some suitable conditions,it changes the surface structure of the polyester, so that the materials surface is filled with many hydrophilic / lipophilic groups.

The contact angle on material of water or oil become bigger, if the surface energy is lower, the hydrophobicity and oleophobicity of the materials is stronger. The ability of material surface depends on the surface morphology and the chemical composition of outer layer; it has nothing to its internal composition and molecular arrangement. Therefore, to study amphiphobic materials (90°<θ<120°),the surface chemical structure and surface morphology must be proceed from synthesis of low surface energy materials.It is found that silicone and fluoride materials have lower surface energy of all substances in nature, the surface energy of fluoride containing materials is about 0.01N/m lower than the silicone containing materials. Fluoride containing materials have a unique performance compared to other materials, so it is the first choice to prepare super hydrophobic materials.

Fluorine is the most negative element, the bond energy of the C-F can be very high, so the fluorine atom existed in fluorosilane can significantly reduce the surface energy of the surface. Fluorine silicon materials has low surface energy for its fluorine and silicon atoms, the Si-F bond energy is very high so it is hardly to be polarized, because of the existence of fluorine atoms. Fluoride silicon materials have a higher solubility in various solvents, because of the presence of silicon atoms, the modified materials have a good flexibility. Fluorine silicon material is suitable for the preparation of super hydrophobic fiber surface.

In this paper, the surface modification of polyester fibers was carried out by chemical grafting method. The effect of modification was represented in two aspects: graft ratio and contact angle. The reaction mechanism and the influencing factors were studied. The optimum reaction conditions were determined. The graft ratio is the relative ratio of the weight of the fiber before and after grafting, which reflects the effect of the grafting. Contact angle reflects the surface tension and the hydrophobicity of fiber surface.

In the experiment, the molecular formula of the fluorine silicon is C8F17(CH2)3Si(OMe)3, and the chemical bonding was formed between the hydrolysate of fluorine silicon and the -OH exposed on the substrate. Reaction mechanism is:

$$C_8F_{17}(CH_2)_3Si(OMe)_3 + 3H_2O \longrightarrow C_8F_{17}(CH_2)_3Si(OH)_3 + Me(OH)_3$$

IOP Conf. Series: Materials Science and Engineering **350** (2018) 012006 doi:10.1088/1757-899X/350/1/012006

2. Experimental

Measurement SEM (Magellan TM 400L XHR), contact angle measuring device (Shanghai Fang Rui, JYC-4), electronic balance (SHIMADAZU)

Reagent polyester fiber filter ball, isopropanol, acetic acid, caustic soda (AR), perfluorooctyltrimethoxysilane (FAS-17, made in our laboratory), KH-570 (AR), DN-101 (AR), peanut oil (commercially available)

Hydrolysis of FAS-17 1g FAS-17 was placed in a conical flask with isopropanol and distilled water, the mole ratio of FAS-17 and distilled water is 1:3, then 100μ L acetic acid was added as the catalyst, standing 24h at room temperature. Remove the solvent and the hydrolysis of FAS-17 was obtained.

Reaction between FAS-17 and polyester *fiber* The polyester fiber ball is placed in the 500mL four-neck boiling flask, and the 200mL distilled water was added and boiling 2h to remove the oil. Be dried 12h in the oven at the temperature of 80°C to constant weight. The constant weight of the polyester fiber ball was described as m1.

200mL isopropanol, FAS-17 hydrolysate and treated polyester fiber ball was added in another 500mL four-hole boiling flask. Reacting at 60°C in a certain time, 100mL×2 of isopropanol was used to wash the polyester fiber ball, take the ball out and put it in oven at the temperature of 80℃ for drying 12h to constant weight, The constant weight of the polyester fiber ball was described as m2. Graft ratio = (m2-m1) /m1.

After the modification, the polyester fiber ball was disassembled to fine fiber wires, aluminum plates were placed above and below the wires, and 10kg weight of heavy stuff was placed above the upper aluminum plate for 24h to form a thin slice. The surface contact angle of the slice was measured at five different points; the final data is the average of five points.

3. Results and discussion

Influence of polyester fiber ball quality to modification effect The concentration of silane coupling agent FAS-17 was unchanged (c=1.9%), different quality of fiber balls were treated at 60℃, and the results were shown in table 1.

Table 1. The graft ratio and the contact angle of the treated fiber wires at 60℃

Time(h)	m1(g)	m2(g)	Graft ratio (%)	Contact angle(°)
1	0.113	0.126	11.50	137
2	0.164	0.184	12.20	137
5	0.124	0.143	15.32	138
9	0.117	0.139	18.80	141
13	0.162	0.195	20.37	144
24	0.158	0.194	22.78	145
30	0.136	0.168	23.53	145

From table 1, it can be seen that under the condition of constant temperature and the certain concentration of coupling agent, the grafting rate increases with time, and the influence of the fiber ball size is not obvious.

In order to easily discuss the relationship between the graft ratio, time (t) and concentration (c), k was introduced, k= (c×t) /m1. If the graft ratio was proportional to the t and c, and reciprocal to the m1, the k value should have a positive correlation with the graft ratio of the fiber. Experimental results are shown in table 2.

Table 2. The data of fiber ball treated with caustic soda and FAS-17hydrolysate

(60℃, 1%NaOH, 1h, c=1.1%)

Time(h)	m1(g)	m2(g)	Graft ratio (%)	Contact angle(°)	k
5	0.193	0. 216	11.92	129	0.284
9	0.123	0.139	13.01	132	0.804
13	0.112	0.127	13.40	134	1.276
18	0.127	0.145	14.17	136	1.559
24	0.109	0.126	15.60	138	2.422
30	0.119	0.138	16.00	140	2.773
34	0.122	0.142	16.40	140	3.065

After surface modification, the contact angle changes with the increasing graft ratio. The grafting ratio increased, the contact angle increased to a certain value, and it could be obtained by the most reasonable grafting rate. The maximum contact angle can reach 145° ,meanwhile the minimum grafting ratio is 23%.

Influence of FAS-17 coupling agent concentration to the modification effect In order to study the effect of silane coupling agent concentration on grafting ratio, the same mass of polyester fiber was taken, and the concentration of the silane coupling agent was changed. Changed the coupling agent dose, the grafting ratio of different concentrations can be obtained, as shown in table 3.

Table 3. Influence of FAS-17 coupling agent concentration to the modification effect(60℃)

Concentration (%)	m1(g)	m2(g)	Graft ratio (%)	Contact angle(°)
1.1	0.331	0.331	0	0
1.4	0.285	0.295	3.51	125
1.7	0.293	0.329	12.28	130
1.9	0.341	0.419	22.87	145
2.0	0.235	0.296	25.95	144
2.1	0.254	0.333	31.10	145
2.2	0.221	0.293	32.57	145
2.5	0.271	0.408	50.55	144
2.8	0.265	0.472	84.37	146
3.1	0.275	0.706	156.72	145

The grafting ratio increased with the increase of the concentration of the silane coupling agent, but the grafting ratio increased rapidly when the concentration over 2%, and the grafting ratio was over 100% at the concentration of 3.1%, the reason for this may be that the silane coupling agent with hydroxyl at the end is very high, which is easy to combine with the surface of the fiber, and the grafting ratio increases with the increase of concentration. But the number of hydroxyl groups of the coupling agent is not determined, and the number of hydroxyl groups that can be combined with the fiber is not determined. With the increase of concentration, the coupling agent is similar to the self coupling reaction, and the coupling of the coupling agent to the surface of the fiber.

The experimental results showed that the saturation grafting ratio of polyester fiber is 23%,and the contact angle is 145°.The water contact angle of the modified fiber did not increase significantly with the increaseing grafting ratio,so it is not necessary to pursue higher grafting rate. After different

IOP Conf. Series: Materials Science and Engineering **350** (2018) 012006 doi:10.1088/1757-899X/350/1/012006

concentrations of modification with FAS-17, the images of the liquid drop on the fiber surface were shown in figure 1.

(a)Water drop, c=1.4% (b)Water drop, c=1.9% (c)Peanut oil, c=1.9%

Figure 1. After modification of FAS-17, the image of the liquid drop on fiber surface

Other coupling agents Two other kinds of coupling agent, silane coupling agent KH-570 and titanate coupling agent DN-101were chosen to replace FAS-17 and reacted with polyester fiber under the same conditions.

Reaction of KH-570 and DN-101 Titanate coupling agent formula can be expressed as RO-Ti (OX-R'Y')n, it is the tetravalent titanium as the center, organic groups linked to the surface of filter material, making the nonpolarity of filter surface enhanced, which is conducive to the treatment of oily wastewater. In short, it is a multifunctional and agile additive. Compared with traditional silane coupling agent, titanate coupling agent is suitable for a wide range, and have a low price.

Titanate coupling agent DN-101 molecular formula is $C_{55}H_{111}O_9Ti$. The modification mechanism of polyester fiber filter surface can be described as: the alkoxy of DN-101 molecule is easy to be hydrolyzed and reacts with the polyester fiber surface hydroxyl groups to form chemical bonds, then the filter surface covered with a layer of DN-101 molecular. And at the other end of the titanate coupling agent molecular, there are three organic macromolecular chain structural units, so oleophilicity and hydrophobicity of the fiber ball filter surface have a great improvement.After modification, the grafting ratio is 1.16%, and the water contact angle of the filter material can reach 141°.

After modification of KH-570 and DN-101, the images of water drops on the fiber surface were shown in figure 2.

KH-570 DN-101

Figure 2. The image of the water drops on fiber surface (modification with KH-570 or DN-101)

4. Conclusions

Isopropanol / H_2O as the solvents, anhydrous acetic acid as the catalyst, perfluorinatedoctylsilane coupling agent FAS-17 can be fully hydrolyzed at room temperature. Studies have shown that perfluorinatedoctylsilane coupling agent FAS-17 is beneficial to gain the hydro-oleophobicity for polyester fiber filter ball. At the optimized conditions of fiber ball modification, the water contact angle of polyester fiber filter ball can reached 145°, and the peanut oil contact angle was 118°.

Several other kinds of coupling agents were used in the same conditions, KH-570 and DN-101 modification can achieved the same hydrophobicity as FAS-17 for polyester fiber ball. These can be used for the filtration of oily wastewater processing, but cannot achieve the same amphiphobic effect compared with FAS-17. Therefore perfluorinatedoctylsilane coupling agent FAS-17 is one more ideal surface modification agent in the field of polymer flooding in oilfield, depth-type filtration of military and aerospace aviation oil, etc.

5. References

[1] Tie Zihan, Wang Xiangcheng, Li Dan etc., Research status and process of polyester fiber surface modification [J]. Chemical Industry times, 2015, 29(8):42-45.

[2] Kota Kitamura, ChihiroOshimo, GeneiMastawa, FutoshiIshimaru. Hindered amine compound, resin composition, Polyurethane Fiber and Production Method and Use thereof: US, 6486238[P], 2002-12-26.

[3] NORIHIRO INAGAKI. Surface modification of PET films by a combination of vinylphthalimide deposition and Ar plasma irradiation [J]. Journal of Adhesion Science and Technology, 2004, 18(13):1517-1528

[4] XuShaogang, Yang Wantai. Study on the mechanism of lepe-AA vapor-phase photograftingsystem[J]. ActaPolymericaSinica, 2000,1(5):594-598.

[5] A. L. Cimecioglu, S. H. Zeronian, K. W. Alger, M. J. Collins, G. C. East. Properties of oligomers present in (ethylene terephthalate) [J]. Journal of Applied Polymer Science, 2003, 32(4):4719 - 4733.

Acknowledgements

The authors want to thank Prof. Jun-he Yang of Shanghai University for providing study conditions.

Microstructure and Corrosion Behavior of Laser Synthesized Cobalt Based Powder on Ti-6Al-4V

O S Adesina[1], A P I Popoola[1], S L Pityana[2] and D T Oloruntoba[3]

[1] Department of Chemical and Metallurgical Engineering, Tshwane University of Technology, P.M.B. X680, Pretoria 0001, South Africa.
[2] Council for scientific and industrial research – National laser Centre, P.O. BOX 395, BLD 46F, Pretoria 0001, South Africa.
[3] Department of Metallurgical and Materials Engineering, Federal University of Technology, Akure, P.M.B. 704, Akure, Ondo - State, Nigeria.
Email: osaadesina@yahoo.com

Abstract. The corrosion behavior of titanium alloys when used for various dynamic offshore components has been a major concern of titanium drilling risers in deepwater energy extraction. A way of achieving specified requirement is the development of coatings suitable to protect the base material against corrosion. In this work, laser cladding technique which is known as a leading edge due to its distinctive properties and outcomes was used in synthesizing Co-based powder on titanium alloy. The processing parameters used were laser power of 900W; scan speed of 0.6 to 1.2 m/min; powderfeedrate1.0g/min; beamspotsize3mm; gasflowrate1.2L/min. The effects of cobalt addition and laser parameters on corrosion behavior of laser clad Ti6AL4V coating in 0.5M sulfuric medium were investigated using linear potentiodynamic polarization. The changes in microstructure and corrosion behavior were analyzed using scanning electron microscopy (SEM) while the X –ray diffraction (XRD) indicates the intermetallics in the coatings. Results showed that the coatings displayed good metallurgical bonding with dendritic formations between the coatings and the substrate. The anodic current density increased with lower scan speed. However, the corrosion current densities of laser-clad samples were lower than Ti6Al4V alloy.

1. Introduction

Since titanium and its alloys became available in the mid-twentieth century, its level of use in offshore and petrochemical applications competes with about 80% of its usage in the aerospace industry [1]. The unique properties of titanium and its alloy such as low density, excellent combination of high specific strength ratio which is maintained at elevated temperature, low modulus of elasticity, good compatibility and good corrosion resistance, makes it a choice material in a wide range engineering application such as power generation, offshore and chemical industries [2, 3]. However, passive behavior of Ti6Al4V in sulphuric, and hydrochloric acids has been a prime concern by many investigators[4, 5]. Contu et al [11] explained that when titanium alloy is submerged in sulphuric acid during applications, it is covered by a thin film of TiH_2. According to Tiyyagura et al [6], hydride formation on the surface is usually not stable for lower over-potentials. In order to mitigate this challenge, there is a need to produce a surface that is deficient of such passive film layer [7]. Surface modification techniques such as laser cladding technique have been employed to fabricate coatings with advanced mechanical and corrosion propertiesin metal surface improvement[8]. Cobalt based alloys tends to be a less deleterious material used mostly as a supporting material with numerous applications in the engineering field because of its excellent resistance to corrosion. Retorts and gas turbine engines or parts (vanes, blades, turbines and compressor discs) are few of the sections in heat treatment plants where cobalt-alloys are commonly utilized and these sections are exposed to long term alternation stresses at elevated temperatures, due to their exceptional strength at high temperatures and also, low susceptibility to sulfidation, nitriding and oxidation [9, 10]. Moreover, the addition of Ni [11] in laser cladding of Co is expected to produce complex phases of CoNi intermetallic with a wide range of superior and functional properties. In this study, a detailed investigation has been undertaken to identify the effect of cobalt based alloy and parameters of laser surface cladding on the microstructure, improvement in electro - chemical behavior of Ti - 6Al - 4V.

2. Experimental

2.1. Materials and coating preparation

In this study, Ti-6Al-4V plates were used as substrates (50 x 50 x 5 mm). The substrate were sandblasted by spraying them with SiO_2 grit sand and cleaned with acetone before exposure to laser beam to influence absorptivity of the laser beam radiation. LSC experiment was carried out using a continuous wave 4.4 kW RofinSinar Nd: YAG laser. The processing parameters used were laser power of 900W; scan speed of 0.6to1.2m/min; powder federate 1.0g/min; beam spot size 3mm; gas flow rate 1.2L/min. In the course of laser cladding process, the molten pool was protected by argon gas flowing at 3 L/min. This was done in order to prevent oxidation as Ti has high affinity for oxygen.

2.2. Microstructural characterization and corrosion measurement

The microstructures of the laser clad coatings were analyzed by SEM equipped with an (EDS) Energy Dispersive Spectroscopy (FE-SEM JSM-7600F) while the phases were examined using Philips PW1713 X-ray diffractometer (XRD) fitted with monochromatic CuKα radiation set at 40Kv and 20mA, while phase identification was done using Philips Analytical X'Pert High Scores software fitted with an in built (ICSD) database. The electrochemical behavior of thespecimen was studied 0.5 M H_2SO_4environment at room temperature using the potentiodynamic polarization method at scan rate of 2mV/S starting from -1.0 V (with respect to the OCP) toabout 1.2 V. Platinum rod was used as the counter electrode and a silver/silver chloride (3 MKCl) as the reference electrode (SSE). The electrochemical parameters were calculated by analyses of the Tafel region using the General Purpose Electrochemical Software (GPES) version 4.9.

3. Results and analyses

3.1. Microstructure and characterization of laser coatings

Figure 1 reveals the SEM image showing the morphologies of the coatings and distributions for 70Co-30Ni coating fabricated at different scan speeds. It is noted that all the coatings revealed good dense and strong metallurgical bonding. This is substantiated by the fact that coatings fabricated by laser cladding technique on titanium (Ti6Al4V) is supported by the action of higher heat input as stated by Tan et al [12]. Figure 1 (a) is characterized by larger grains while Figure. 1(b) is characterized by smaller grains which could be seen at top right of the coating. At Figure 1 (a) lower scan speed revealed grain growth with equiaxed structures. The rationale is that, at low laser scan speeds, the clad layer rests on a larger dilution zone, due to a prolonged energy input from the heat source to support inter-diffusion with migration of elemental atoms between the coating and the substrate [13]. The coating in Figure 1b was exposed to rapid heating and solidification owing to high scanning speed, therefore resulting in uneven heating rate which lead to transformation from equiaxed to dendritic microstructures structure. Unlike the equiaxed grains associated with coating deposited at lower scan speed, Figure 1b is characterized by a mixture of thin columnar grains and dendritic structures owing to increased solidification rate at higher scan speed [14]. Higher scan speed is characterized by increased cooling rate as a result of rapid heat transport from the melt pool leading to formation of larger band of dendritic structures. Literature has shown that rapid solidification of Co tends to grow as dendrites [15].

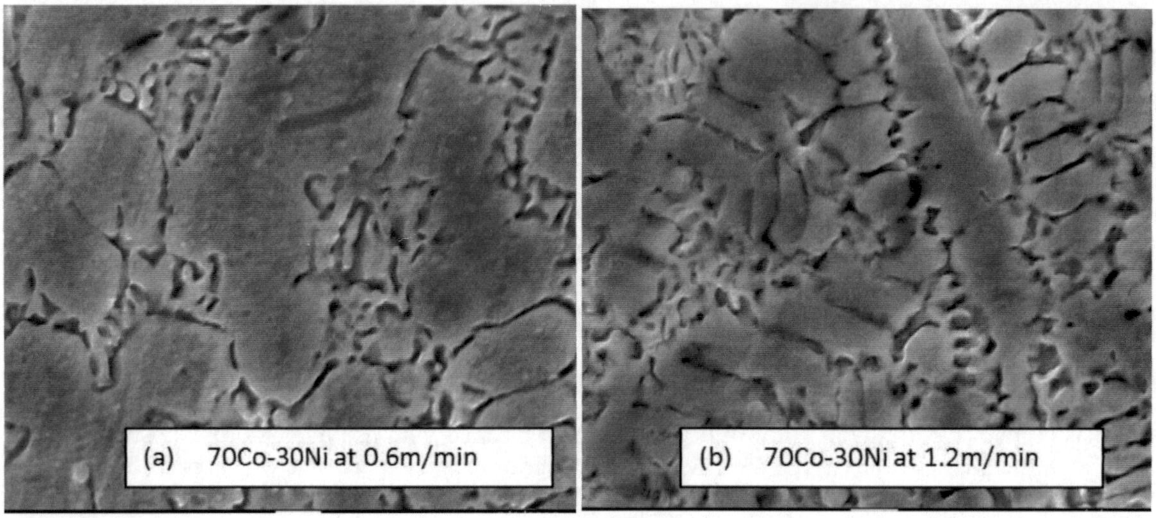

Figure 1: SEM showing microstructural evolution of 70Co-30Ni clad samples at (a) 0.6 m/min and (b) 1.2 m/min at 900 W.

3.2. Phase analysis

The XRD patterns displayed in Figure 2 clearly shows that the phases formed in these coatings differ significantly, as a result of varying laser parameters. At 0.6m/min scan speed the phases revealed Ti_2Ni, Co_3Ti, $Al_{13}Co_4$ while at 1.2 m/min, the XRD spectrum result revealed additional novel phases of NiTi and Ni_3Ti. Intermetallic phases of Co_3Ti and $Al_{13}Co_4$ were identified as dominant phases in the spectrum. Weng et al [16] stated that high energy radiated by the laser beam lead to molten powders interactions with substrate, thereby yielding complex chemical reactions. Consequently, emanation of Ti and other alloying elements from the substrate at both scan speeds indicates sufficient energy intensity to encourage intense material diffusion and mixing between the substrate and the coating [14].

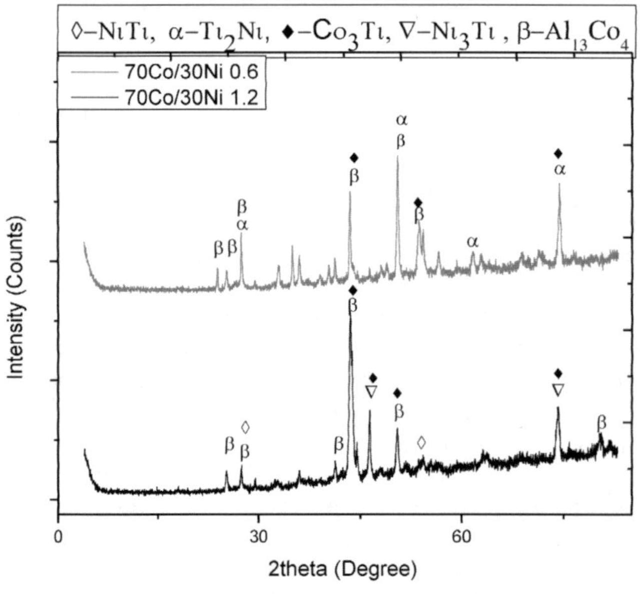

Figure 2: X-ray diffraction analysis of laser Clad 70Co-30Ni clad samples.

3.3. Corrosion behavior

Figure 3 illustrates the potentiodynamic polarization curves of 70Co/30Ni laser cladded on Ti6Al4V substrate at scan speeds 0.6 and 1.2 m/min both at laser power of 900W. Electrochemical oxidation behaviors were investigated with the help of potentiodynamic polarization technique in 0.5M H_2SO_4. An increase in laser scan speed resulted in corrosion current density of the fabricated coatings declining from 0.0000473mA to 0.0000417mA, thus revealing the effect of laser speed on corrosion current by changing the grain size of 70Co-30Ni alloy. There is a slight passivation trend over the range of potentials for the substrate on the anodic branch. However, Ti6Al4V is known to form protective film on the alloy surface which turns to slow down/stop accelerated electrochemical surface deterioration [17]. The anti-corrosion of fabricated coatings may be linked to simultaneous effects of important factors, for instance chemical composition, phase structure, grain size and preferred orientation [18]. Non-uniform distribution of the Co-Ti reinforcement can be observed on the microstructure of this coating on both 0.6m/min and 1.2m/min samples. The grain boundaries show no sign of micro pores, no cracks and intermediate homogeneity of the fabricated is displayed. 70Co/30Ni at 0.6m/min is represented by Ti_2Ni, $Al_{13}Co_4$ and Co_3Ti phases as shown in Figure 2. Hence, improved corrosion properties are displayed by the fabricated coatings. This is due to the fact that Ni-Ti phase is nobler than Ti6Al4V [19]. High laser scan speed (1.2m/min) resulted in less potential value while low scan rate (0.6m/min) resulted in increased potential for coating. The 70Co-30Ni coating at 0.6 m/min displays a slight primary passivation potential while 1.2 m/min 70Co-30Ni fully shows passivation with a greater passivation area and a prolonged trans-passive region.

Table 1: Potentiodynamic polarization data of samples in 0.5M H_2SO_4

Laser parameters	Ecorr, Calc (V)	Ecoor, Obs (V)	icorr (A/cm^2)	CR (mm/yr)	Rp ($\Omega.cm^2$)
Control	-0.24539	-0.26757	0.00016067	1.3855	627.35
70Co/30Ni- 0.6 (900)	0.033378	-0.039468	4.73E-05	0.40819	1455
70Co/30Ni- 1.2 (900)	-0.18729	-0.22669	4.17E-05	0.36001	777.86

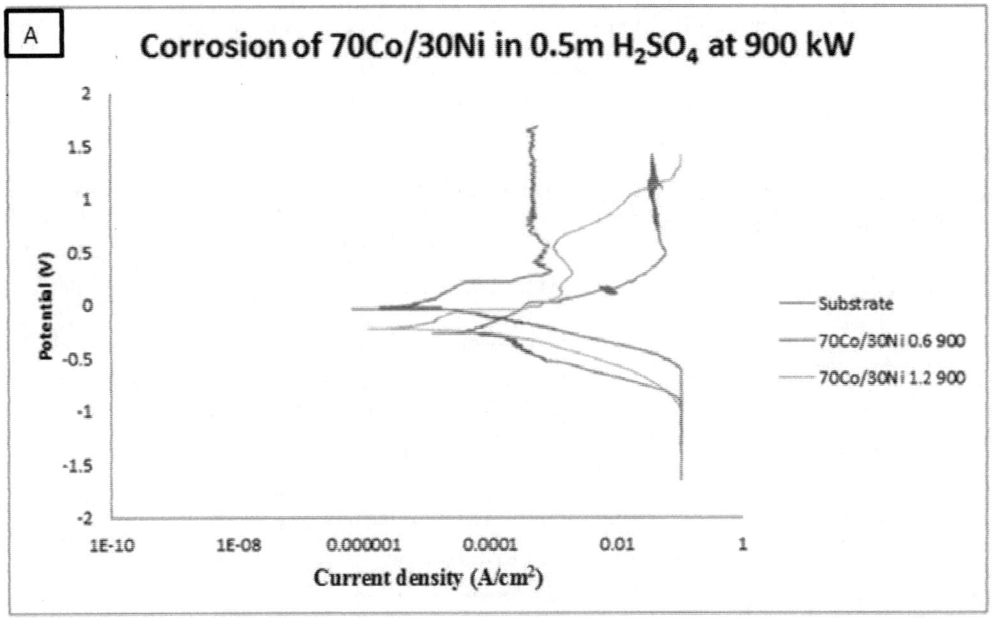

Figure 3: Potentiodynamic polarization curves of Ti6Al4V substrate and laser clad 70Co/30Ni at 0.6m/min and 1.2m/min.

IOP Conf. Series: Materials Science and Engineering **350** (2018) 012007 doi:10.1088/1757-899X/350/1/012007

4. Conclusion
Coatings produced a crack free, dense and strong metallurgical bond with significant effect on the microstructure and corrosion properties. The microstructural characteristic at lower scan speed shows larger grains with equiaxed structures while high scan speed is characterized by a mixture of thin columnar grains and dendritic structures owing to increased solidification rate. XRD analysis confirms the presence of novel phases such as Ti_2Ni, Co_3Ti, and $Al_{13}Co_4$. These intermetallic phases dispersed within the coating matrix could be responsible for the improved corrosion obtained. Laser clad coatings exhibited active-passive behavior in sulfuric environment. The anodic current density increased with lower scan speed. Likewise, the corrosion current densities of laser-clad samples were lower than Ti6Al4V alloy. In summary, cobalt is considered a more active element in improving the corrosion behavior of titanium in aggressive corrosive medium.

5. Reference
[1] Bloyce A, Morton P and Bell T 1994 *ASM handbook* **5** 835-851.
[2] Baker T N 2010 *Surface Engineering of Light Alloys* 398-443.
[3] Chikarakara E, Naher S and Brabazon D 2012 *Surf. Coat. Technol* **206** 3223-3229
[4] Oliveira V, Aguiar C, Vazquez A, Robin A, and Barboza M 2014 Corros. Sci. **88** 317-327.
[5] Contu F, Elsener B, and Böhni H, 2004 Corros. Sci **46** 2241-2254.
[6] Tiyyagura H and Kumar K 2017 *Journal of Materials Sci & Surface Engi.* **5 (1)** 496 - 499
[7] ASM International, 2003. *Corrosion.* **13**A.
[8] Obadele B, Lepule M, Andrews A and Olubambi P 2014 *Tribo.Intern.* **78** 160-167.
[9] Zhang K, Liu M.M, Liu S.L, Sun C and Wang H .F 2011 *Corros. Sci* **53** 1990-1998.
[10] Kim K, Lee S, Nam N and Kim J 2011 *Corros. Sci*, **53** 3576-3587.
[11] Zhang Y, Tu Y, Xi M and Shi L 2008 *Surf. Coat. Technol* **202** 5924-5928.
[12] Tan Y F, He L, Wang X L, HongH O N G and Wang W G 2014 *Transactions of Nonferrous Metals Society of China* **24** 2566-2573.
[13] Mahamood R M, Akinlabi E T, Shukla M and Pityana S 2013 *Mater. Des.* **50** 656-666.
[14] ShujaS Z and Yilbas B S 2011*Optics & Laser Technology* **43** 767-775.
[15] Frenk A, Henchoz N, and Kurz W 1993 *Z.Metallkd.* **84** 886-892.
[16] Weng F, Yu H, Chen C and Dai J 2015 *Mater. Des.* **80** 174-181.
[17] Hu L F, Li J, Tao Y F and Lv Y H 2017*Surf. Coat. Technol.* **311** 295-306.
[18] Bakhit B, Akbari A, Nasirpouri F and Hosseini M G 2014*Appl. Surf. Sci.* **307** 351-359.
[19] Mokgalaka M N, Pityana S L, Popoola P A I and Mathebula T 2014 *Adv. in Mat. Sci & Engi*

Acknowledgment
This research is supported by the National Laser Centre (NLC), Council of scientific and industrial research (CSIR), Pretoria, South Africa.

IOP Conf. Series: Materials Science and Engineering **350** (2018) 012008 doi:10.1088/1757-899X/350/1/012008

Study of Interaction between Cadmium and Bovine Serum Albumin with UV-Vis Spectrocopy Approach

E Suhartono[1], I Thalib[2], I Aflanie[3], Z Noor[4] and R Idroes[5]

1 Department of Medical Chemistry/Biochemistry, Faculty of Medicine, Lambung Mangkurat University, Banjarmasin 70712, South Kalimantan, Indonesia
2 Research and Development Unit, Mutiara Bunda Mother and Child Hospital, Martapura 70614, South Kalimantan, Indonesia
3 Department of Forensic, Faculty of Medicine, Lambung Mangkurat University, Banjarmasin 70232, South Kalimantan, Indonesia
4 Department of Orthopaedic Surgery, Faculty of Medicine, Lambung Mangkurat University, Banjarmasin, South Kalimantan, Indonesia
5 Department of Chemistry/Pharmacy, Faculty of Mathematics and Natural Sciences, Syiah Kuala University, Aceh 24415, Aceh, Indonesia
E-mail: esuhartono@unlam.ac.id

Abstract. This study aims to explain the interaction of cadmium (Cd) with serum albumin through visible light (UV-Vis) spectroscopy approach. This study is an in vitro experimental study using Cd with several concentrations and Bovine Serum Albumin (BSA). Each solution was then incubated for 10 min at 37°C, and measured the absorbance at 220-300 nm. The absorbance data is then presented in graphical form. From the graph, a linear equation will appear to calculate the value of metal binding constants (K) to proteins. Also, in this present study we analsyed the ratio between A220 and A220 to identify changes in the protein region especially tyrosine and peptide bonds. The results show that the addition of Cd in different concentrations could increase the absorbance with a constant value (K) = 1.634. Based on the result, it seems the addition of Cd in different concentrations will lead the reaction to form BSA-Cd. Also, the result shows that the ration of A220/A280 were decreased with the increasing of Cd concentration. In conclusion, the addition of Cd could interact and changes the protein structure in BSA.

1. Introduction

Cadmium (Cd) used since 1950 with a total world production of about 15,000-18,000 tons per year. Cd is a heavy metal that is widely used by humans for the benefit of various industries, for example agricultural industry, paint, aircraft, and others. In addition to beneficial, Cd is also harmful to the environment and adversely affect human health, such as hepatic, ovarian, kidney, and brain damage [1]-[6]. Previous research has also revealed that Cd can lead to increased blood cholesterol and rat endothelial damage [7].

Cd ability to cause oxidative damage and metabolic disorder begins with the binding of Cd covalently with functional groups of amino acids making up the protein, for example -OH, -COO-, -C = O, -SH, -S-S-, -NH2 and -NH. The binding of the Cd with the functional group may form a crosslink of the residue of ditirosin and disulfide bonds in the protein. The cross linking is called Advanced Oxidation Protein Products (AOPPs). Increased AOPP results in changes in protein structure and function [8]-[9]. The results of Arizal et al. [6] have concluded that the administration of Cd could significantly increase the levels of AOPP in rats ovarium, resulting in impaired ovarian function. Meanwhile, the results Aflanie et al. [10] also concluded that the increased concentration of Cd will be followed by the increasing level of AOPPs in vitro. Nevertheless, in those studies, there has been no explanation of Cd interactions with amino acids that cause crosslinking of ditirosin and disulfide.

Many methods are used to explain metal interactions with protein. However, the spectroscopic method is a frequently used method. One of them is visible light spectroscopy (UV-Vis). The basic principle of UV-Vis spectroscopy is that the molecule has an electron energy level analogous to the level of electron energy in the atom. The energy level of this molecule is called a molecular orbital. Thus, any change in molecular structure will give different uptake. Zhang et al. [11] had studied spectroscopically the interaction of flavonoid metal divalent Cu2+ and protein. In these study, it was revealed that Cu2+ interacts with flavonoids to form flavonoids-Cu2+ and a change in protein conformation due to flavonoids-Cu2+. Curvale et al. [12] studies have examined the interaction of Arsenic (As) with amino acids in proteins by UV-Vis spectroscopy [12]. In that study, it was revealed that As ionic charge can affect the interaction of its bonding to proteins. The interaction between As (III) and protein is an ionic interaction, but As (V) interaction is a hydrophobic interaction.

The use of bovine serum albumin (BSA) as an interaction model between metal and protein is due to its ability to reversibly bind various ligands. Albumin contains 17 disulfide bridges and thiol groups which are free cysteine (Cys). This residue (Cys 34) and tyrosine (Tyr) are reactive, polarized, and have a high affinity for metals and are very sensitive to the local environment [12]. In this regard, in order to explain the interaction of Cd with protein, a research between Cd in various concentrations with the bovine serum BSA through spectroscopic approach should be done for a deeper understanding.

2. Material and methods

2.1. Research design

This study is an experimental study to explain the interaction of metals and proteins. Thus, the following groups were created: P0 = BSA; P1 = BSA + 0.001 ppm Cd^{2+}; P2 = BSA + 0.01 ppm Cd^{2+}; P3 = BSA + 0.1 ppm Cd^{2+}. Each group was incubated for 10 min at 37°C. Then, in each group sample the absorbance was measured at a wavelength of 220-300 nm. The absorbance data will be presented in graphical form.

2.2. Spectroscopic analysis

Each solution from each group was diluted with normal saline at concentration of 5 μL/mL. Then, the spectra were scanned in the wavelength 220 and 280 nm. Amide chain of proteins present in the blood absorbs strongly at 220 nm, and absorption of tyrosine and tryptophan is at around 280 nm. Then, the ratio between A220 and A280 was analsyed which is the result of absorbance at 220 and 280 wavelength in each experimental group.

2.3. Data analysis

The interaction between the metal (L) with protein (P) is based on the reaction equation [13]:

Metals + Proteins ⇆ Metals : Protein

So the bonding constant (K) is formulated:

$$K = \frac{[\text{Metal:Protein}]}{[\text{Metal}][\text{Protein}]} \qquad (1)$$

If it is assumed [Metals: Protein] = C_b, then

$$K = \frac{[\text{Cb}]}{[\text{Cl}-\text{Cb}][\text{Cp}-\text{Cb}]} \qquad (2)$$

Under Lambert-Beer's law, then

$$Cl = \frac{\text{Ao}}{\varepsilon\text{l.b}}$$

$$Cp = \frac{\text{Ao}}{\varepsilon\text{p.b}}$$

By entering a value of C_l and C_p in equation (2) then obtained equation (3):

$$\frac{Ao}{A - Ao} = \frac{\varepsilon\text{l}}{\varepsilon\text{b}} + \frac{\varepsilon\text{l}}{\varepsilon\text{b.K}} \cdot \frac{1}{Cl}$$

If n = εb/εl, then the equation (4) is as follows:

$$\frac{1}{A - Ao} = \frac{1}{\text{n}} + \frac{1}{\text{n.K}} \cdot \frac{1}{Cl}$$

By using graph linear curve between the $\frac{1}{A-Ao}$ and $1/C_L$ will be obtained the values of n, the number of the active site binding proteins and metal on metal-protein binding constants (K).

3. Results and discussion

Based on the observations, the addition of Cd concentration causes an increase in absorbance. This shows that there is an interaction between Cd and BSA, as shown in Figure 1.

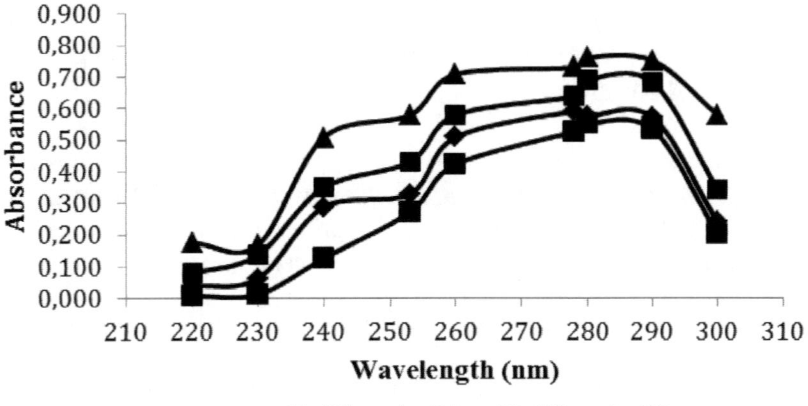

Figure 1. Graph of wavelength relation with absorbance. P0 = BSA; P1 = BSA + 0.001 ppm Cd^{2+}; P2 = BSA + 0.01 ppm Cd^{2+}; P3 = BSA + 0.1 ppm Cd^{2+}

Figure 1 shows that an increase in Cd concentration causes absorbance band change at 220-300 nm wavelength. The alteration is thought to be due to the covalent bonding of Cd with the N-terminal amide group, the N-terminal of the imidazole group, and the N-terminal of the amine group. This results in changes in the structure of protein molecules, which lead to changes in absorbance bands.

The results are consistent with research Dieaconu et al. [14] that was indicated that the metal binding can affect the absorbance band at the BSA by UV-Vis spectroscopy. This study is also consistent with the results of research Curvale et al. [12]. In these study, it was revealed that BSA absorbance band-As^{5+} higher than the BSA-As^{3+}. This implies that the binding interactions BSA metal-As $^{5+}$ more powerful than the BSA-As^{3+} at a wavelength of 200-300 nm.

Many studies have revealed the association of changes in absorbance bands with some diseases. Kumar et al. research, [15] has identified absorbance bands in IgG of Duchene Muscular Dystrophy (DMD) patients using UV-Vis spectroscopy. In these study, it was found that in patients with DMD the absorbance of protein were decreased. Research of Gunasekaran et al. [16] also reveals that the UV-Vis absorbance band wer found lower in jaundice patient's than normal in the 200-250 nm regions. Some researchers have used the A280/A220 ratio as a parameter to identify changes in the protein region. It is based on that A280 is the absorbance of tyrosine and tryptophan residues that are sensitive

to change. Meanwhile, A220 is the absorbance of protein-peptide bonds. Changes in the A280/A220 ratio mean that protein structures undergo many changes, especially in the amino acid tyrosine and peptide bonds (figure 2) [17].
In this study also measured the ratio of A280 / A220 to know the molecular changes of protein region. The calculation results are presented in Figure 2.

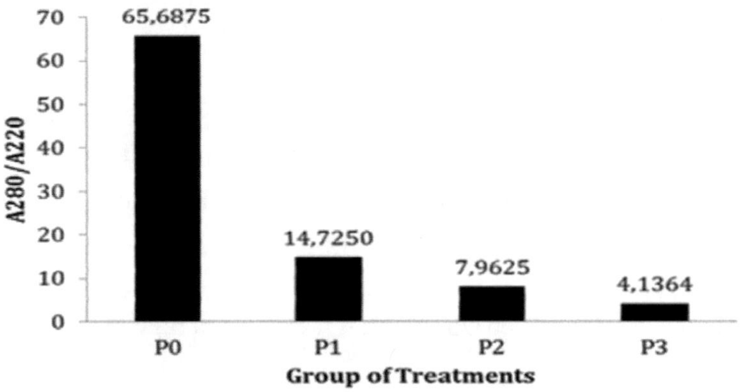

Figure 2. Ratio of A280 / A220 in different group of treatments. P0 = BSA; P1 = BSA + 0.001 ppm Cd^{2+}; P2 = BSA + 0.01 ppm Cd^{2+}; P3 = BSA + 0.1 ppm Cd^{2+}

Figure 3 shows that the increasing of Cd concentration will decrease the A280/A220 ratio. By many researchers, the use of A280/A220 ratio has been used for the identification of several diseases, such as cancer and inflammation [19]. This is evidenced by research Gunasekaran et al that the ratio of A278/A210 leukemia patients were significantly decreased compared to healthy people [19]. In addition, the study also revealed that A278/A210 blood of healthy people is higher than those suffering from anemia and hepatic sirosis.
In addition to calculating the ratio of A280/A220, this study also measures the value of the BSA-Cd dictation constant. The graph of the measurement results can be seen in figure 3.
Based on graph 3, we can calculate the equilibrium constant value of BSA-Cd binding based on equation formula (4). The calculation result shows that the constant of BSA-Cd dictation is K = 1,634. This value means that the reaction between BSA and Cd tends towards the formation of BSA-Cd.

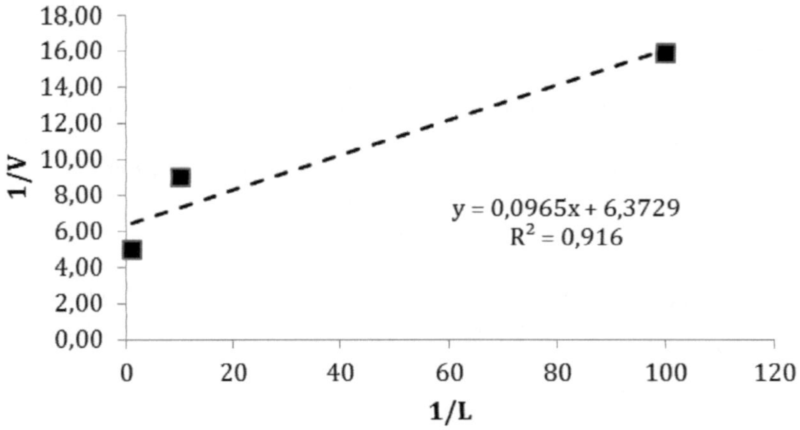

Figure 3. Graphs between 1 / V with 1 / L to determine the value of K

4. Conclusion

Based on the results of the study, it was concluded that the addition of Cd in different concentrations will increase the abosorbance which indicates the covalent binding of Cd in the amino acid protein constituents. The lower A280 / A220 ratio means that the protein structure changes a lot. Meanwhile, the BSA-Cd binding constant value is 1.634.

5. References

[1] Suhartono E, Triawanti, Yunanto A, Firdaus RT, and Iskandar 2013 Chronic cadmium hepatooxidative in rats: Treatment with Haruan fish (*Channa striata*) extract. *APCBEE Procedia.* **5** 441-5.

[2] Suhartono E, Nijka JA, Anhar VY, Sari RA, Edyson, and Marisa D 2015 Anti-Lipid peroxidation activities of three selected fruits juices against cadmium induced liver cells damage in vitro. *J. Trop. Life. Sci.* **5**, 75-9.

[3] Suhartono E, Triawanti, Leksono AS, and Djati MS 2014 Oxidative Stress and kidney glycation in rats exposed to cadmium *Int. J. Chem. Eng. App.* **5** 497-501.

[4] Suhartono E, Triawanti, Leksono AS, and Djati MS 2015 Effects of cadmium exposure on lipid peroxidation and chlorinative stress of rat kidney. *J. Exp. Life. Sci.* **5** 1-5.

[5] Suhartono E, Iskandar, Hamidah S, and Arifin YF 2015 Phytochemical constituents analysis and neuroprotective effect of leaves of *N. coriacea* (*Nothaphoebe Coriacea*) on cadmium-induced neurotoxicity in rats: An in-vitro study *Int. J. Toxicol. Pharm. Res.* **7** 297-302.

[6] Ramadhani EA, Suhartono E, and Noor MS 2014 The difference of *Advanced Oxidation Protein Products* (AOPPs) level in rats (*Rattus novergicus*) ovarian exposed and not exposed with cadmium *Berkala Kedokteran.* **10** 144-7.

[7] Anindya WIA, Muhyi R, and Suhartono E 2016 Coronary hearth disease risk induced by cadmium through the measurement of cholesterol level and *circulating endothelial cells* in plasma of rats. *Berkala Kedokteran.* **12** 153-63.

[8] Witko-Sarsak V, Friedlander M, Khoa TN, Capeillere-Blandin C, Nguyen AT, Canteloup J, Drayer JM, Jungers P, Drueke T, and Deschamps-Latscha B 1998 Advanced oxidation protein products as novel mediators of inflammation and monocyte activation in chronic renal failure. *J. Immunol.* **161** 2524–32.

[9] Veyseller B, Aksoy F, Ertas B, Keskin M, Ozturan O, Yildirim YS, Bayraktar FGI, and Ozturk H 2010 A new oxidative stress marker in patients with nasal polyposis: advanced oxidation protein products (AOPP). *Acta. Otorhinolaryngol. Belg.* **6** 105-9.

[10] Aflanie I, Muhyi R, Suhartono E 2015 Effect of heavy metal on Malondialdehyde and advanced oxidation protein products concentration: A focus on arsenic, cadmium and mercury. *J. Med. Bioeng.* **4** 332-7.

[11] Zhang Y, Shi S, Sun X, Xiong X, Peng M 2011 The effect of Cu^{2+} on interaction between flavonoids with different C-ring substituents and bovine serum albumin: Structure–affinity relationship aspect. *J. Inorg. Biochem.* **105** 1529–37.

[12] Curvale RA, Debattista NB, Pappano NB, Gonzalez DM, Curvale DA 2011 Spectrophotometric study of the interaction between arsenic and bovine serum albumin. *Res. Reprod. Biol.* **6** 131-5.

[13] Kanakis CD, Tarantilis PA, Polissiou MG, Diamantoglou S, Tajmir-Riahi HA 2006 Antioxidant flavonoids bind human serum albumin. *J. Mo. Struc.* **798** 69–74.

[14] Dieaconu M, Ioanid A, Iftimie S, Antohe S 2012 UV-Absorption mechanisms of Ni-binding bovine serum albumin. *Dig. J. Nanomater. Biostruc.* **7** 1125-38.

[15] Kumar S, Sweety, Chaudhary S 2015 Study of immunoglobulin 'G' with ultra violet spectroscopy in Duchene muscular dystrophy and Alzheimer's disease. *J. Biomed. Eng. Med. Imag.* **1** 51-79.

[16] Gunasekaran S, Uthra D 2008 FTIR and UV-Visible spectral study on normal and jaundice blood Samples *Asian. J. Chem.* **20** 5695-703.

[17] Gunasekaran S, Natarajan RK, Renganayaki V, Rathikha R 2008 FTIR and UV visible spectrophotometric approach to discriminate leukemic sera. *Asian. J. Chem.* **20** 2521-30.

[18] Dieaconu M, Ioanid A, Iftimie S, Antohe S 2012 UV-Absorption mechanism of Ni^{2+}- binding bovine serum albumin. *Digest. J. Nanomater. Biostruc.* **7** 1125-38.

[19] Gunasekaran S, Natarajan RK, Renganayaki V 2008 UV visible spectrophotometric approach and absorption model for the discrimination of disease blood. *Asian. J. Chem.* **20** 48-54.

Acknowledgments

The authors are grateful to the financial support of Faculty of Medicine, Lambung Mangkurat University in funding this project through the Faculty Research grant.

Interaction betwen Lead and Bone Protein to Affect Bone Calcium Level Using UV-Vis Spectroscopy

Z Noor[1], A Azharuddin[2], I Aflanie[3], N Kania[4], and E Suhartono[5]

1 Department of Orthopaedic Surgery, Ulin General Hospital/Faculty of Medicine, Lambung Mangkurat University, Banjarmasin, South Kalimantan 70232, Indonesia
2 Department of Orthopaedic Surgery, Faculty of Medicine, Syiah Kuala University, Aceh 24415, Indonesia
3 Department of Forensic, Ulin General Hospital/Faculty of Medicine, Lambung Mangkurat University, Banjarmasin, South Kalimantan 70232, Indonesia
4 Research Center for Toxicology, Cancer, and Degenerative Disease, Department of Pathology, Ulin General Hospital/Faculty of Medicine, Lambung Mangkurat University, Banjarmasin 70232, South Kalimantan, Indonesia
5 Department of Medical Chemistry/Biochemistry, Faculty of Medicine, Lambung Mangkurat University, Banjarbaru, South Kalimantan 70712, Indonesia
E-mail: esuhartono@unlam.ac.id

Abstract. This present study aim to evaluate the interactions between lead (Pb) and with bone protein by UV-Vis approach. In addition, this prsent study also aim to investigate the effect of Pb on bone calcium (Ca) level. The present study was a true experimental study design to examine the impact of Pb exposure in bone of male rats (*Rattus novergicus*). The study involved 5 groups, P1 was the control group, while the other (P2-P5) were the case group with exposure of Pb in different concentration within 4 weeks. At the end of the exposure, the interaction between Pb and protein was determined using UV-Vis spectrophotometric method, and the Ca level was determined using permanganometric method. The results shows that that there is an interaction between Pb and bone protein. The result also shows that the value of the binding constant of Protein-Pb is 32.71. It means Pb have an high affinity to bind with bone protein, which promote a further reaction to induced the release of bone Ca from the bone protein. In conclusion, this present study found an obvious relationship between Pb and bone protein which promote a further reaction to increase the releasing of bone calcium.

1. Introduction

Lead (Pb) is a heavy metal belonging to period IV of the periodic table. This element is grey in color, malleable, soft, and can be extracted from the mineral galena (PbS). Pb is beneficial to humans for industrial purposes such as battery industry, X-ray protection, paint coloring, PVC industry and others. Widespread use of Pb, causing this element to be a pollutant to the environment and harmful to human health [1]. Pb can enter the body through the digestive tract, breathing, and skin. Pb entering through the gastrointestinal tract will be distributed to bone (60%), liver (25%), kidney (4%), reticuloendothelial system (3%), intestinal wall (3%), and to other tissues. The accumulation of Pb in these organs can result in various organ disorders, such as impaired kidney function, cardiovascular, liver, brain, and hematopoietic system [2].

Disorders of various organs due to Pb are caused by the interaction of Pb with various cations, especially those of the divalent cation, such as calcium (Ca), zinc, and iron [3]. Pathological processes can be found in cell membranes and mitochondria, in the synthesis and function of neurotransmitters, heme synthesis, cellular redox status, and nucleotide metabolism. In addition, Pb also interacts with sulfihydryl groups, carboxyl groups, amine groups of enzyme amino acids, cell membrane constituents, and other macromolecules [4].

Side effects of Pb have been shown in many previous studies. The study by Monir et al. [5] proved that Pb may decrease collagen and increase the risk of fracture in adult female C57 / B6 rats given drinks containing Pb for 4 months. A study in the Hezhang region of China found an association between Pb levels in the blood and a decrease in BMD in women [6]. Furthermore, a study conducted by Todorovic et al. [7] revealed that mice exposed at 100 mg Pb^{2+} per kg of body weight for 30 days may have a decrease in bone Ca by 16%. This is due to the greater Pb^{2+} affinity than Ca^{2+}, causing Pb^{2+} to replace Ca^{2+} in bone mineral matrix, thus lowering Ca^{2+} bone levels [7].

Many methods are used to explain the interaction of metals with proteins, one of which is visible light spectroscopy (UV-Vis). The basic principle of UV-Vis is that the molecule has an electron energy level analogous to the energy level of the electrons in the atom. The energy level of this molecule is called a molecular orbital. Study on the interaction between metal and protein with the UV-Vis method has been described previously [8]. Zhang et al. [8] investigated the interaction between flavonoid-divalent metal Cu^{2+} and protein. The result revealed that Cu^{2+} interacts with flavonoids to form flavonoids-Cu^{2+} and a change in protein conformation due to flavonoids-Cu^{2+}. Curvale et al. [9] studies have examined the interaction between Arsenic (As) with amino acids in proteins by UV-Vis. The result revealed that the As could interacted with proteins. The interaction between As (III) and protein is an ionic interaction, and between As (V) with the protein is a hydrophobic interaction [9].

However, as far as we know there is a view literature that explains the effect of Pb on bone protein through the UV-Vis approach. Thus, our present study aim to evaluate the interactions between Pb and with bone protein by UV-Vis approach. Furthermore, this study will explain a deeper understanding how the Pb could changes in the bone Ca level.

2. Material and methods

2.1. Study design

This study was of pure experimental post test-control group design, conducted for 4 weeks from April to May 2017.

2.2. Experimental section

Male albino rats (*Rattus norvegicus*) aged 2-3 months, weighing 80-200 grams, obtained from BPPV, Banjarbaru were used in this study. Before intervention, the rats were acclimatized for 7 days, and kept in separate cages according to their assigned group. During the acclimatization period the animals were given exactly the same animal feed pellets and tap water as were to be used during the intervention. Immediately before the intervention, the rats were fasted for 1-2 hours to ensure that their proventriculus was empty.

Furthermore, the subjects were divided into 5 groups, P0 = control groups; P1 = administration of 0.01 mg / L $PbCl_2$; P2 = administration of 0.1 mg / L $PbCl_2$; administration of 1 mg / L $PbCl_2$; administration of 10 mg / L $PbCl_2$; administration of 100 mg / L $PbCl_2$. Selection of the animals was by simple random sampling, while the sample size was calculated by means of Federer's formula, yielding a sample size of 5 rats per group. After treatment, the rats og each group was killed with ether and their bones taken for determination of calcium content by the permanganometric method, as performed in a previous study [10].

2.3. Bone collection
The collected bones were cleaned and washed, then left to dry in an oven. The dried bones were then pulverized and passed through a 40 mesh sieve. The bone powder was dissolved in HCl (1:4), the resulting solution was concentrated by evaporation and then placed in a water bath for 1 hour. The dry residue was redissolved in 5 ml concentrated HCl and 50 ml distilled water, then again placed in the water bath for several minutes, and filtered through #52 Whatman filter paper.

The filtrate was collected in a 200 ml volumetric flask. The precipitate still remaining on the filter paper was washed in distilled water and the wash water was added to the collected filtrate. Distilled water was added to the mixture of filtrate and wash water to yield a volume of 200 ml. This solution was labeled aliquot A.

2.4. Calcium concentration determination
A volume of aliquot A equivalent to 2 g bone ash was pipetted into a 300 ml beaker and diluted to 200 ml with distilled water. The solution was made slightly alkaline by the addition of NH_4OH (1:4) using methyl orange as indicator. HCl (1:4) was then added to slightly acidify the solution, after which 10 ml 0.5 N HCl and 10 ml 2.5 % oxalic acid solutions were added, and the mixture heated to a boil. Under continuous stirring, 15 ml saturated ammonium-oxalate solution was added. The mixture was further heated until a granular precipitate was formed, then the mixture was left to cool. 8 ml of 20% sodium acetate was added and the mixture left for 12 hours. Next the mixture was filtered and washed with hot water until free of chloride (washed with hot water and then with a small amount of HCl (1:4) and finally with hot water until free of chloride). The residue was then put into a beaker by perforating the tip of the filter paper cone with a glass stirrer, then sprinkled with just sufficient hot water until all precipitate had fallen into the beaker. Next, 10 ml H_2SO_4 (1:4) was added and the dissolved precipitate heated to near-boiling, left to cool, then titrated with 0.1 N $KMnO_4$. The calcium content was calculated using the formula:

$$Ca\ per\ 100\ mg\ sample = \left(\frac{ml\ titration\ \times total\ volume\ of\ dissolved\ ash}{volume\ of\ dissolved\ ash\ \times weight\ of\ sample}\right) \times 100\%$$

2.5. UV-Vis spectroscopy determination
Aliquot A is taken 100 mL and then put into a cuvette. Absorbance was observed at wavelengths of 220-300 nm using a spectrophotometer.

2.6. Metal-Protein bonding constant determination
Metal-protein binding constants were calculated using formula:

$$\frac{1}{A - Ao} = \frac{1}{n} + \frac{1}{n.K} \cdot \frac{1}{Cl}$$

Using a linear curve graph between 1/(A-Ao) with 1/Cl, we get the n value, ie the number of active sites that bind metals to proteins and Metal-Protein (K) binding constants.

2.7. Ethical clearance
The present study was approved by the Reaserch Ethics Commission of the Faculty of Medicine, Lambung Mangkurat University.

3. Results and discussion
Based on the observations, the addition of Pb concentration causes an increase in absorbance at 280 nm wavelength. This shows that there is an interaction between Pb and Tyr, as shown in Figure 1.

Figure 1 shows that an increase in Pb concentration causes the absorbance of the Protein-Pb binding to increase. This is thought to be due to the covalent binding of Pb with the N-terminal amide group, N-terminal of the imidazole group, and the N-terminal of the amine group. This results in changes in the structure of protein molecules [10]. The binding constant of Protein-Pb is determined using the graph as in Figure 2.

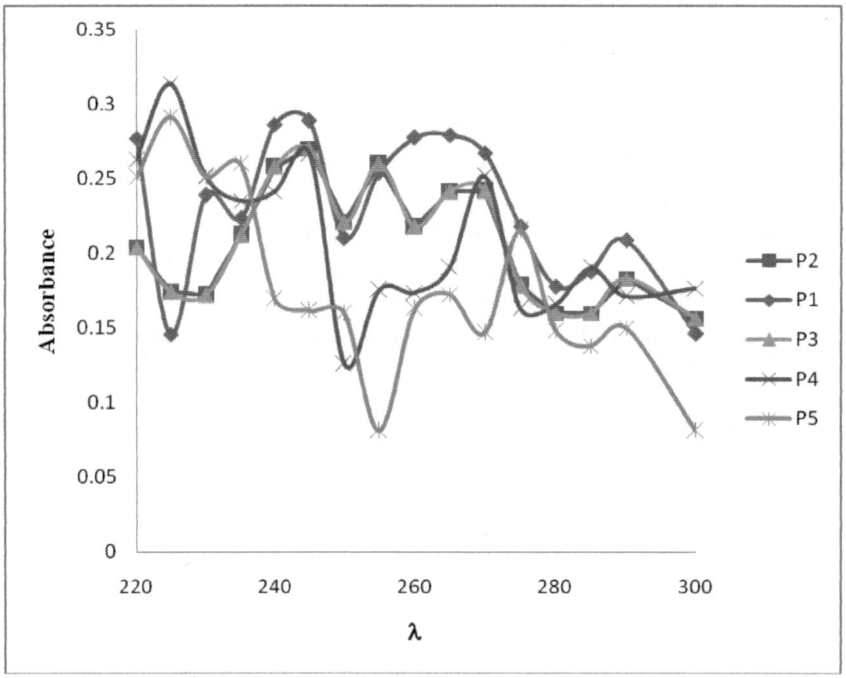

Figure 1. UV-Vis spectroscopy analysis of bone in different group of treatments. P1 represent control group; P2 represent the 0.01 mg/l $PbCl_2$ treatment group; P3 represent the 0.1 mg/l $PbCl_2$ treatment group; P4 represent the 1 mg/l $PbCl_2$ treatment group; and P5 represent the 10 mg/l $PbCl_2$ treatment group.

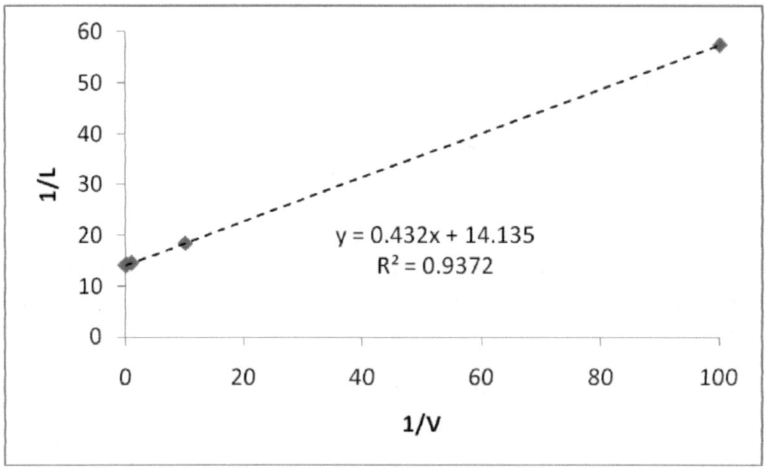

Figure 2. The linear curve to determine metal-protein binding constant (K)

According to Figure 2, the value of the binding constant of Protein-Pb is 32.71. The positive value of K means that the reaction between Pb and the protein will shift to the right. This means that the affinity between protein and Pb is increase, so from this point of view we can said that Pb have a tendency to bind bone protein forming Protein-Pb. The bond between Pb and bone proteins will then lead to the release of bone calcium [11]. This can be seen from the results of this study which is presented in Figure 3.

According to the Figure 3, the level of bone calcium is decrease with the increasing of Pb exposure. It means, the greater the level of Pb exposure, the more calcium will be released from the bone. However, the exact mechanism how the Pb could affected the bone calcium level via the interaction with protein is still unclear. Our previous study indicated that kinetically Pb and calcium has a similar property. Pb and calcium both are divalent cation. However, since Pb has more a much higher affinity than calcium, the presence of Pb in bone could displace calcium from the bone [12].

Furthermore, how the Pb could displace bone calcium, and how is the interaction mechanism between Pb and bone protein still can not be known clearly in this research. It is a limitation of this study that can be used as a reference for further research.

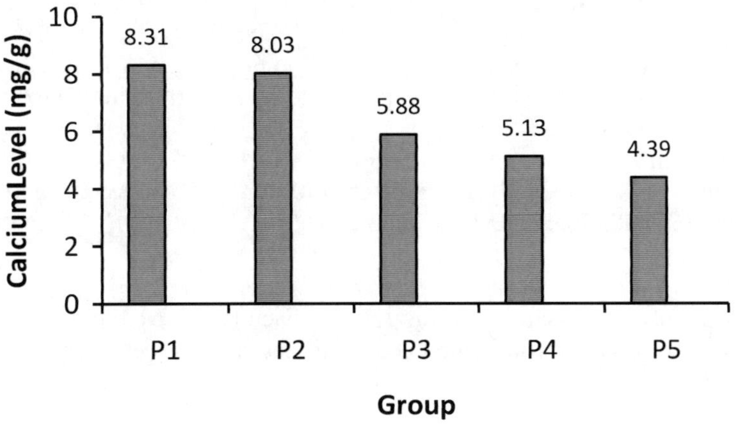

Figure 3. Bone calcium level in different group of treatments. Values are mean of three replicates in each group of treatment. P1 represent control group; P2 represent the 0.01 mg/l $PbCl_2$ treatment group; P3 represent the 0.1 mg/l $PbCl_2$ treatment group; P4 represent the 1 mg/l $PbCl_2$ treatment group; and P5 represent the 10 mg/l $PbCl_2$ treatment group.

4. Conclusion

In conclusion, this present study found an obvious relationship between Pb and bone protein which promote a further reaction to increase the releasing of bone calcium. These findings suggest that many efforts should be made to reduce environmental Pb exposures.

5. References

[1] Leonas R, Noor Z, Rasyid HN, Madjid TM, and Tanjung FA 2017 Effect of lead nanoparticles inhalation on mesostructure and the osteoprotegerin/receptor activator of nuclear factor-kappaB ligand system in rats. *Cukurova. Med. J.* **41** 48-54

[2] Zang X, Yang L, Li Y, Li H, Wang W, and Ye B 2012 Impacts of lead/zinc mining and smelting on the environment and human health in China *Environ. Monit. Assess.* **184** 2261-73

[3] Ani M, Moshtaghie AA, and Aghadavood M 2006 Protective effects of selenium and zinc on the brain acetyl cholinesterase activity in lead intoxified rat *Res. Pharm. Sci.* **2** 80-4.

[4] Kasperczyk S, Blaszczyk I, Dobrakowski M, Romuk E, Kapka-Skrzypczak L, Adamek M, and Birkner E 2013 Exposure to lead affects male biothiols metabolism. *Annals. Agric. Environ. Med.* **20** 721–5.

[5] Monir AU, Gundberg CM, Yagerman SE, van der Meulen MCH, Budell WC, Boskey AL, and Dowd TL 2010 The effect of lead on bone mineral properties from female adult C57/BL6 mice *Bone.* **47** 888-94.

[6] Chen X, Wang K, Wang Z, Gan C, He P, and Liang Y 2014 Effects of lead and cadmium co-exposure on bone mineral density in a Chinese population *Bone.* **63** 76-80.

[7] Todorovic T, Vujanovic D, Dozic I, and Petkovic-Curcin A 2008 Calcium and magnesium content in hard tissues of rats under condition of subchronic lead intoxication. *Magnesium. Res.* **21** 43-50.

[8] Zhang Y, Shi S, Sun X, Xiong X, and Peng M 2011 The effect of Cu^{2+} on interaction between flavonoids with different C-ring substituents and bovine serum albumin: Structure–affinity relationship aspect. *J. Inorg. Biochem.* **105** 1529–37.

[9] Curvale RA, Debattista NB, Pappano NB, Gonzalez DM, and Curvale DA 2011 Spectrophotometric study of the interaction between arsenic and bovine serum albumin *Res. Reprod. Biol.* **6** 131-5

[10] Suhartono E, Ulfarini YW, Triawanti, Mustaqim WA, and Firdaus RT, and Setiawan MHM 2012 Increase bone calcium dissociation in lead-exposed rats. *Univ. Med.* **31** 151-8.

[11] Haleagrahara N, Chakravarthi S, Kulur AB, and Radhakrishnan A 2011 Effects of chronic lead acetate exposure on bone marrow lipid peroxidation and antioxidant enzyme activities in rats *Afr. J. Pharm. Pharmaco.* **5** 923-9.

[12] Lu H, Yuan G, Yin Z, Dai S, Jia R, Xu J, Song X, Li L, and Lu C 2014 Effects of subchronic exposure to lead acetate and cadmium chloride on rat's bone: Ca and Pi contents, bone density, and hitopathological evaluation. *Int. J. Clin. Exp. Pathol.* **7** 640-7.

Acknowledgments
The authors are grateful to the financial support of Faculty of Medicine, Lambung Mangkurat University in funding this project through the Faculty Research grant.

IOP Conf. Series: Materials Science and Engineering **350** (2018) 012010 doi:10.1088/1757-899X/350/1/012010

Designing a Polymerase Chain Reaction Device Working with Radiation and Convection Heat Transfer

M Madadelahi [1*], K Kalan [2*], A Shamloo[3]

1 Research Assistant, School of Mechanical Engineering, Sharif University of Technology
2 BSc Student, School of Mechanical Engineering, Sharif University of Technology
3 Associate Professor, School of Mechanical Engineering, Sharif University of Technology
E-mail: shamloo@sharif.edu

Abstract. Gene proliferation is vital for infectious and genetic diseases diagnosis from a blood sample, even before birth. In addition, DNA sequencing, genetic finger-print analyzing, and genetic mutation detecting can be mentioned as other procedures requiring gene reproduction. Polymerase chain reaction, briefly known as PCR, is a convenient and effective way to accomplish this task; where the DNA containing sample faces three temperature phases alternatively. These phases are known as denaturation, annealing, and elongation/extension which in this study -regarding the type of the primers and the target DNA sequence- are set to occur at 95, 58, and 72 degrees of Celsius. In this study, a PCR device has been designed and fabricated which uses radiation and convection heat transfer at the same time to set and control the mentioned thermal sections. A 300W incandescent light bulb able to immediately turn off and on along with two 8×8 cm DC fans, controlled by a microcontroller as well as PID and PD controller codes are used to monitor the applied thermal cycles. In designing the controller codes it has been concerned that they not only control the temperature over the set-points as well as possible, but also increase the temperature variation rate between each two phases. The temperature data were plotted and DNA samples were used to assess the device function.

1. Introduction

In 2001 Celera Corporation along with scientists from China, USA, Japan, Germany, France, and Britain published a draft of human genome [1], which was a landmark in anthropology science. Their results demonstrated that 99.99% of the genes were identical among all humans and only one-in-millions differences made the vast variety of people. Thus, discovering these tiny differences required vast investigation in medical and biological fields.

PCR, or polymerase chain reaction, is a technique through which a specific sequence of DNA can multiply for millions of times during a short period. Inventing this method in 1983 brought the Nobel Prize for Kary Mullis [2] because of being cheaper and more effective than the former expensive and time consuming conventional methods. Nowadays, PCR is widely used in laboratories for diverse applications such as genes identifying, separating, and cloning as well as living beings classification and identification, genetic diseases diagnosis, even criminal cases. PCR comprises three main sections namely denaturation, annealing, and elongation/extension occurring at 90-95, 50-65, and 70-74 degrees of Celsius, respectively. During denaturation stage, the hydrogen bonds between complementary bases

[*] Equal contribution

break; consequently, the double-stranded DNA turns into two separated strands. In the next step, annealing, the primers stick to the single-stranded DNA templates. The next temperature stage, called elongation or extension, is the stage in which Taq-polymerase enzyme synthesizes new complementary strands to the already present and separated strands using free nucleotides suspended inside the sample. Stationary PCR -a well-known, conventional, effective, easy, and cheap sort of the PCR devices- is now widely being used and studies have been done to improve its functionality [3-7]. In this class of PCR devices, the DNA containing sample is steady and by utilizing equipment, temperature varies among the set-points. Therefore, the sample warms up and cools down without moving. Cho et al. [8] developed a rapid real-time micro-scale chip-based PCR system consisting of six individual thermal cycling modules capable of independent control of PCR protocols. They also conducted large-scale clinical evaluation study in order to test utility of a chip-based PCR system as a molecular diagnostic device. Consolandi et al. [9] tested a silicon biochip designed for PCR analysis of complex biological samples. They recorded a cooling rate of 3.2 °C/s and a heating rate of 11 °C/s during thermal cycling. Their device amplified the 2060-bp cyanobacterial 16S rRNA gene and the 330-bp human anti-α_1-chymotrypsin gene. Kim et al. [10] designed a system to measure the temperature distribution inside a silicon PCR micro-channel, where they used a Pt film heater and a Pt film sensor. Bustin [11] investigated on the possibility of reducing PCR cycling time and denaturation temperatures. His results interestingly indicated that it is feasible to obtain good PCR results with short denaturation and annealing times and at low denaturation temperatures, which drastically improves the overall time of the reaction. In this paper, a low cost, light and portable stationary PCR device has been designed and fabricated which, applies radiative and convective heat transfer to control the thermal cycles. Its functionality has been assessed by testing a 324-bp DNA sample.

2. Experimental setup

A 300w incandescent light bulb capable of immediately turning off and on along with two 8×8 cm DC fans were chosen as heating and cooling sources, respectively. A micro-controller was coded by PID and PD codes to control the sources. The micro-controller controlled the system by turning two relays off and on, one connected to the lamp and the other to the fans.

A small container with the shape of a truncated square pyramid with top(larger) and base surfaces of 17×17 and 5×5 cm, respectively, was designed and fabricated to hold and fix the lamp and the sample-holder chip. The burner of the lamp was bolted to the base section of the container and the chip was located on the top section. By doing so, the shape of the container helped conducting the reflected light from the walls to the chip; thereby, increasing the radiative power transmitted to the chip. Additionally, a special polished steel plate was opted as the material of the container. This kind of steel, similar to a mirror, reflected the light very well rather than storing it as its internal energy. Devising these arrangements, we maximized the radiative power transmitted from the lamp to the chip. A 0.5 mm thick polycarbonate plate with a surface area of 17×17 was located on the top section of the container, on which, the sample container vial was fixed. Two fans were placed on a fixture which itself was placed on the top of the container. Hence, the chip was located between the light bulb and the fans, in an open interaction with them. Since a great deal of the output energy of the light bulb is transferred in convection mode and in order to prevent this energy from running out through the walls, an elastomeric insulator was wound around the container. Figure 1 shows a schematic and real view of the temperature control system.

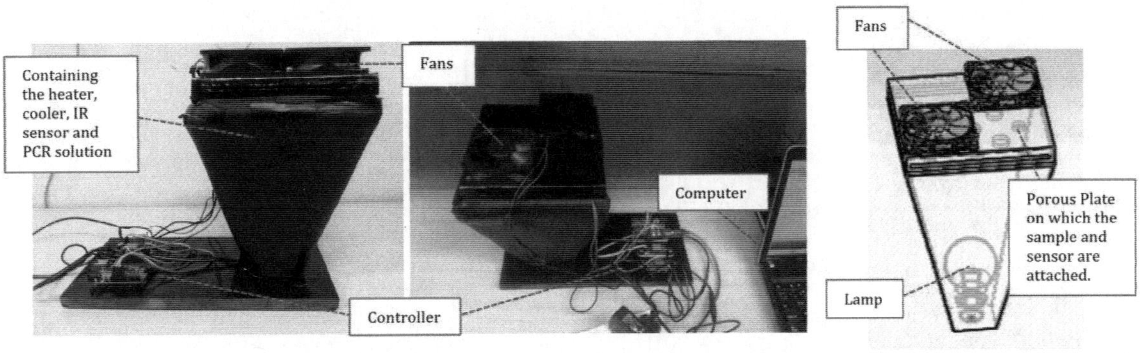

(a) (b)

Figure 1. Temperature control system. (a) real device (b) schematic design.

An infra-red temperature sensor module with an accuracy of ±0.02 degree of Celsius was placed on the chip to read the temperature. In order to evaluate the function of this sensor, a K-type thermocouple with an accuracy of ±0.5 was also connected to the chip and presented the data through a BTM-4208SD both online and saved on a SD card. The thermocouple was placed close to the sensor and the data from the thermocouple and the sensor demonstrated a difference lower than 0.5 degree. The temperature data charts read from the sensor also were both plotted online and offline.

3. Controller and temperature control

As previously mentioned, in a PCR reaction, three temperature sections of denaturation, annealing, and elongation, each for 30 seconds and at temperatures of 90-95, 50-65, and 70-74 degrees of Celsius, must occur alternatively for 34 cycles. Furthermore, system was set to work for 3 minutes once in the beginning and at the denaturation temperature, and once for another 3 minutes at elongation temperature in the end. Codes of the micro-controller were written so that they could fix the temperature of the system around the set-points.

This controller adjusted the temperature by controlling the on and off time of the relays, one connected to the lamp and the other to the fans. Temperature read by the sensor was given to the system as the feedback. To control the system and considering that the on and off time of each source had to change with respect to the error from the set-point, a PID and a PD controller were designed for the system. Each integral, derivative, and proportional coefficient was determined according to the system response and by trial and error.

There were two main ideas to control the system. First, a PID controller was designed that controlled denaturation and elongation stages only by turning the lamp on and off and the annealing stage by turning only the fans on and off. It was expected to achieve a good accuracy at the cost of speed. The PID controller in comparison with the PD code was more time consuming, but more accurate due to the integral proportion. Each cycle using the PID code took about 250 seconds in which, regarding that in each cycle there is only a 90-second-period needed for the stages, about 160 seconds were extra. In other words, the ramp of changing the temperature between each two stages was too slow. The temperature data are plotted in Figure 2.

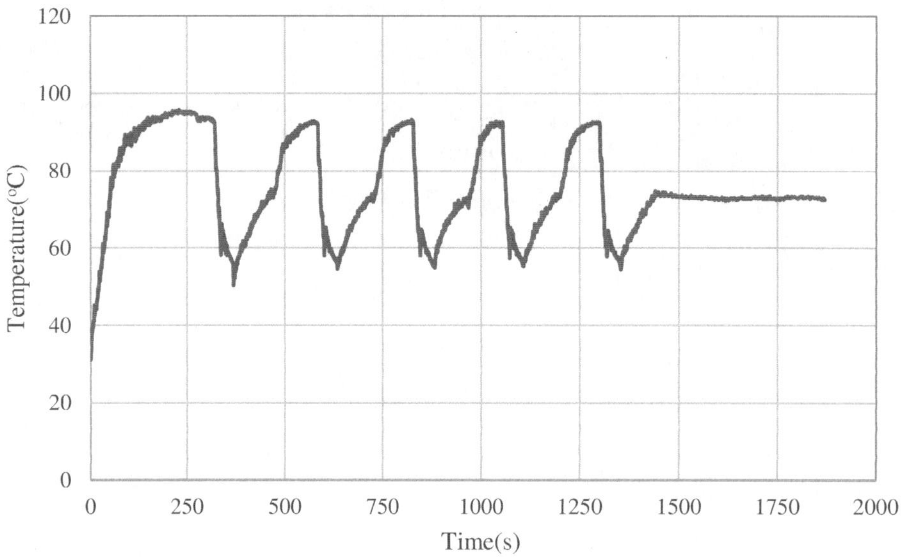

Figure 2. Temperature data of five cycles of the system when using PID controller.

In the second experiment, integral proportion was eliminated from the controller and additionally, each stage was controlled by turning both the heating and cooling sources on and off, where the total time for each cycle reduced to about 160 seconds (Figure 3).

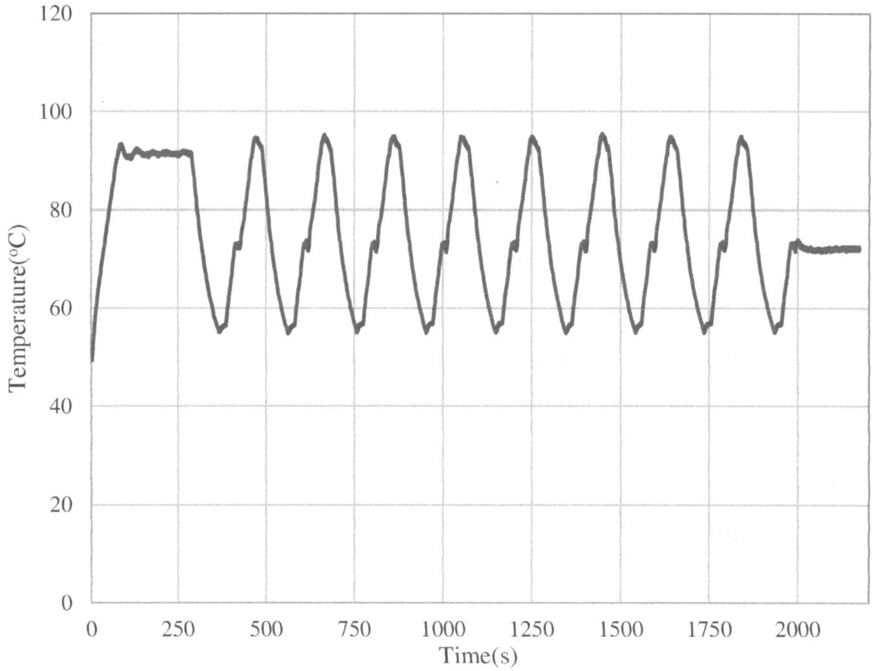

Figure 3. Temperature data of nine cycles of the system when using PD controller.

Here, increasing the ramp of the temperature variation between two stages was the first priority. However, calculation of the standard deviation using for both tests (Table 1 and Table 2), demonstrated slight increase in the error, at the second test.

Table 1. Temperature and standard deviation data of the system controlled by PID controller.

	Denaturation	Annealing	Elongation
Set-point	92.00 °C	58.00 °C	72.00 °C
Average temperature	91.86 °C	57.59 °C	71.37 °C
Standard deviation	0.56	1.55	0.72

Table 2. Temperature and standard deviation data of the system controlled by PD controller.

	Denaturation	Annealing	Elongation
Set-point	92.00 °C	56.00 °C	72.00 °C
Average temperature	92.20 °C	56.25 °C	72.33 °C
Standard deviation	1.05	0.72	0.90

4. Tests and results

A PCR sample consists of different components, such as primers, gene samples, nucleotides, Taq polymerase enzymes, buffers, etc. For this test, a 324 kb DNA was aimed to be multiplied. The sample was prepared and injected into a 0.2 μl vial (Figure 4).

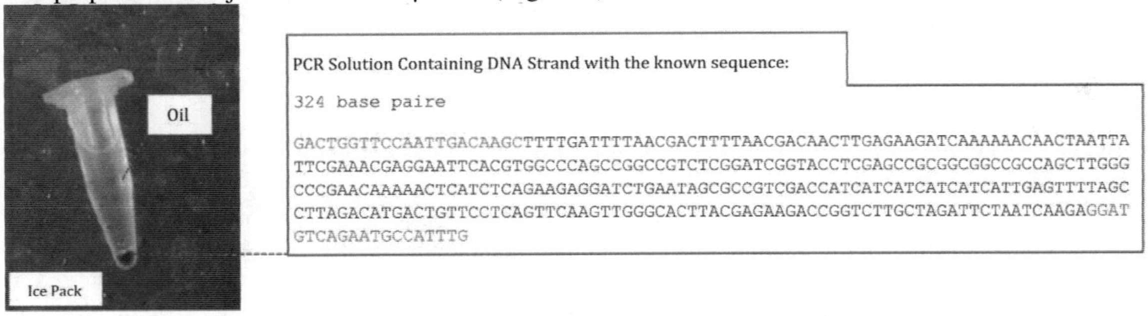

Figure 4. Sample containing vial placed on an ice pack.

After each test, the sample was evaluated using a gel electrophoresis device to measure the quality and quantity of the proliferation. Figure 5 shows the result of the reproduction, when using the PID controller (column 3) and using our commercial PCR device (column 1). As can be seen, there is no bond in the range of the target DNA and the halo represents the dimer among the primers which has been formed due to the unfavorable temperature conditions.

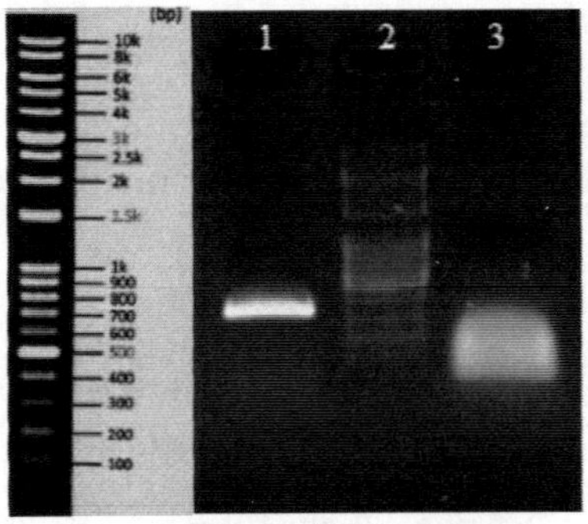

Figure 5. Gel electrophoresis results of the multiplied sample when using PID controller.

Figure 6 represents the result of the gel electrophoresis analysis on the tested sample, when system was using the PD code as its controller. Unlike the previous case, a bright belt in the length range of the target DNA was visible, which means that along with the primer dimers, DNA has been multiplied as well. Although the low brightness of this belt and the primer-primer dimers demonstrate the low quantity and quality of the proliferation, respectively, this device has been accomplished to do the reproduction task.

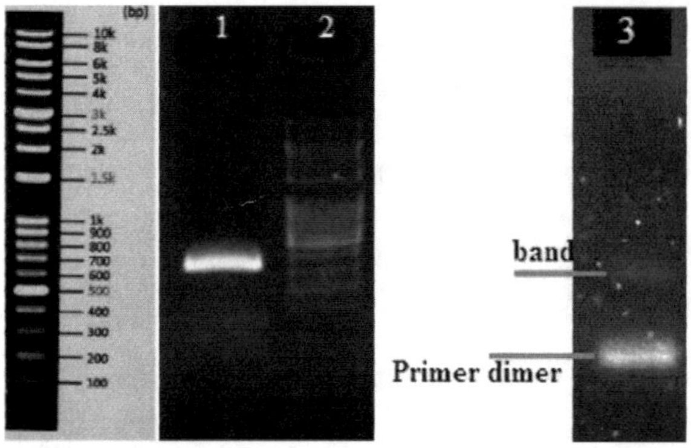

Figure 6. Gel electrophoresis results of the multiplied sample when using PD controller.

Comparing the two tests and their results, it can be concluded that the ramp of the temperature variation is much more important than its accuracy. Reducing the extra time of each cycle drastically improved the functionality of the device.

5. Conclusion

A low-cost, light, and portable PCR device was designed, fabricated, and tested which utilized radiation and convection heat transfer simultaneously, to control each stage. Two controllers were designed for the system, and the functionality of the system using each two was assessed by multiplying a 324-bp DNA and analyzing the reproduction using a gel electrophoresis device. Hereunder comes a brief conclusion:

1. The ramp of the temperature variation is more important than the accuracy of adjusting the temperature. In other words, reducing the extra time of each cycle will result in better proliferation. By reducing the cycle time from 250 to 160 seconds, we could observe the proliferation.
2. As the system uses radiation besides convection for heating the sample, the response time of the sample will be comparatively high; meaning that the difference between the vial wall and sample temperature will be low. On the other hand, as the system uses only forced convection heat transfer and the vial wall is made of plastic which is not a good heat conductor, temperature inside and outside of the vial will be considerable. To eliminate this problem, annealing and elongation time can be raised.
3. Radiation heat transfer can increase the temperature rising ramp; thereby, improving the system functionality.
4. The annealing stage could be considered as the point of design; for removing the energy from the sample is harder than giving internal energy to it.

6. References

[1] Venter, J. Craig, Mark D. Adams, Eugene W. Myers, Peter W. Li, Richard J. Mural, Granger G. Sutton, Hamilton O. Smith et al. "The sequence of the human genome." *science* 291, no. 5507 (2001): 1304-1351.

[2] Mullis, Kary, Fred Faloona, Stephen Scharf, R. K. Saiki, G. T. Horn, and H. Erlich. "Specific enzymatic amplification of DNA in vitro: the polymerase chain reaction." In *Cold Spring Harbor symposia on quantitative biology*, vol. 51, pp. 263-273. Cold Spring Harbor Laboratory Press, 1986.

[3] Kopp, Martin U., Andrew J. De Mello, and Andreas Manz. "Chemical amplification: continuous-flow PCR on a chip." *Science* 280, no. 5366 (1998): 1046-1048.

[4] Obeid, Pierre J., Theodore K. Christopoulos, H. John Crabtree, and Christopher J. Backhouse. "Microfabricated device for DNA and RNA amplification by continuous-flow polymerase chain reaction and reverse transcription-polymerase chain reaction with cycle number selection." *Analytical chemistry* 75, no. 2 (2003): 288-295.

[5] Norian, Haig, Ryan M. Field, Ioannis Kymissis, and Kenneth L. Shepard. "An integrated CMOS quantitative-polymerase-chain-reaction lab-on-chip for point-of-care diagnostics." *Lab on a Chip* 14, no. 20 (2014): 4076-4084.

[6] Sun, Yingnan, Xiaoguang Zhou, and Yude Yu. "A novel picoliter droplet array for parallel real-time polymerase chain reaction based on double-inkjet printing." *Lab on a Chip* 14, no. 18 (2014): 3603-3610.

[7] Nanayakkara, Imaly A., Weidong Cao, and Ian M. White. "Simplifying Nucleic Acid Amplification from Whole Blood with Direct Polymerase Chain Reaction on Chitosan Microparticles." *Analytical Chemistry* 89, no. 6 (2017): 3773-3779.

[8] Cho, Yoon-Kyoung, Jintae Kim, Youngsun Lee, Young-A. Kim, Kak Namkoong, Heekyun Lim, Kwang W. Oh et al. "Clinical evaluation of micro-scale chip-based PCR system for rapid detection of hepatitis B virus." *Biosensors and Bioelectronics* 21, no. 11 (2006): 2161-2169.

[9] Consolandi, Clarissa, Marco Severgnini, Andrea Frosini, Giancarlo Caramenti, Marco De Fazio, Francesco Ferrara, Anna Zocco, Alessandra Fischetti, Michele Palmieri, and Gianluca De Bellis. "Polymerase chain reaction of 2-kb cyanobacterial gene and human anti-α 1-chymotrypsin gene from genomic DNA on the In-Check single-use microfabricated silicon chip." *Analytical biochemistry* 353, no. 2 (2006): 191-197.

[10] Kim, Soo Ho, Jermim Noh, Min Ku Jeon, Ki Woong Kim, Luke P. Lee, and Seong Ihl Woo. "Micro-Raman thermometry for measuring the temperature distribution inside the microchannel of a polymerase chain reaction chip." *Journal of Micromechanics and Microengineering* 16, no. 3 (2006): 526.

[11] Bustin, Stephen A. "How to speed up the polymerase chain reaction." *Biomolecular Detection and Quantification* 12 (2017): 10-14.

IOP Conf. Series: Materials Science and Engineering **350** (2018) 012011 doi:10.1088/1757-899X/350/1/012011

An Approach for Improvement of Carbon Fiber Technique to Study Cardiac Cell Contractility

T Myachina[1], A Khokhlova[1,2], I Antsygin[1], O Lookin[1,2]

1 Ural Federal University, Ekaterinburg, Russia
2 Institute of Immunology and Physiology, Russian Academy of Sciences, Ekaterinburg, Russia
E-mail: a.d.khokhlova@urfu.ru

Abstract. The technologies to study cardiac cell mechanics in near-physiological conditions are limited. Carbon fiber (CF) technique is a unique tool to study single cardiomyocyte contractility. However, the CF adhesion to a cell is limited and it is difficult to control CF sliding occurred due to inappropriate adhesion. In this study, we present a CF adhesion quality index – a linear coefficient (slope) derived from "end-diastolic cell length - end-diastolic sarcomere length" relationship. Potential applicability of this index is demonstrated on isolated rat and guinea pig ventricular cardiomyocytes. Further improvement of the approach may help to increase the quality of the experimental data obtained by CF technique.

1. Introduction

The heart is a mechanically functioning blood pump and many studies on cardiac mechanics are focused on pressure-volume relation parameters. Increased hemodynamics (volume and load) is associated with increased wall tension, stretching cardiac cells and their sarcomeres. Isolated cardiomyocytes are an important object to study cardiac mechanics in details on cellular and subcellular levels, because they are free from connective tissue, which cause significant modification of tissue contractile properties.

Currently, the tools to study cardiomyocytes contractility in near-physiological conditions are limited, and the majority of research on single intact cardiomyocytes is conducted on mechanically unloaded cells [1]. There are few technologies to apply defined mechanical extensions to cells: hypoosmotic stress approach [2], stress applied by a glass stylus [3], cell-in-a-gel approach [4], and carbon fiber (CF) technique [5-7].

Le Guennec and co-workers were among the first to use thin 12 μm diameter CFs that were produced by high temperature treatment of polyacrilonitryl and oxidized to produce free carboxyl groups. These fibers stick to the cell membrane without the use of glue, probably because of microstructures at the surface of CF (Figure 1a and b) that interact with the amino groups of the membrane surface. Note that smooth fibers without microstructures (Figure 1c and d) do not stick to the cardiomyocytes [8]. This approach allows attachment of carbon fibers of known stiffness to both ends of a cardiomyocyte. CF-based stretching gives rise to a relatively homogeneous increase in sarcomere length (SL) and allows calculation of passive and active forces from optically monitored changes in CF bending [9].

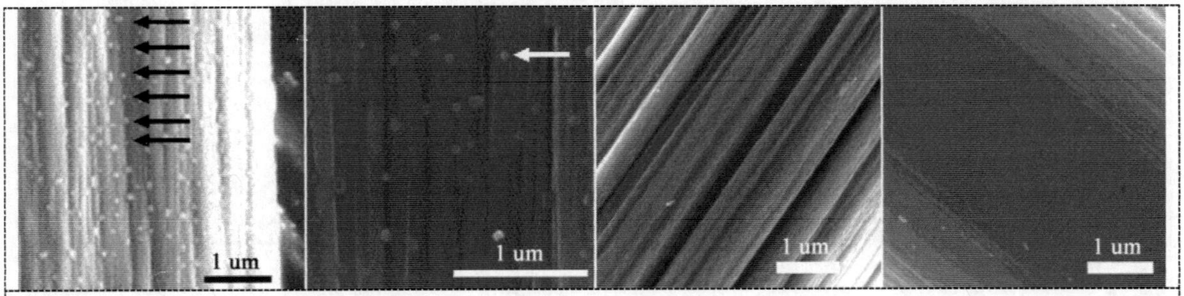

Figure 1. Scanning electron microscopic images of sticky and non-sticky CF, scale bars = 1 μm. (a) Detail of the surface of a sticky CF from Prof. Jean-Yves Le Guennec, (b) The surface of a sticky CF at higher magnification. Arrows show microstructures seen on the fibers, (c and d) non-sticky CF. Adapted from [8].

More recently, to obtain better CF adhesion, graphite-reinforced CFs of higher stiffness were processed by carbonization of rod-shaped graphite granules and resin oligomer mixtures [10]. A higher stiffness allowed recording of higher forces at a small displacements increasing the sensing range.

A modification of CF approach – a new CF 'gripping' technique where both ends of a cardiomyocyte are tightly squeezed between a pair of CFs was recently developed by Iribe and co-workers [6]. This improved technique allows one to increase stretch range in isolated cardiomyocytes from ≈ 4% to 16% in SL compared to conventional CF method.

A common limitation of CF technique is the lack of control of CF sliding relative to the cell surface. Since in the CF method the cell length is determined optically as the distance between the fiber edges the sliding causes incorrect recordings of the stretched cell length that does not correspond to changes in SL estimated using Fourier Transform. In this study, we present an approach to assess CF sliding defining a CF adhesion index. We believe that this method and its further improvement can be used to increase the quality of experimental data obtained by CF technique.

2. Materials and methods

The single myocyte stretch system is described in detail elsewhere [5]. Briefly, thin borosilicate capillaries were pulled from glass tubes and fibers were mounted in the fine tip of the capillaries. The narrow end of the capillary was thermally bent by 45° to allow parallel alignment of fibers with the bottom of the perfusion chamber. The wider end of the glass tube was mounted in the holder of a computer-controlled piezoelectric transducer (PZT) for precise positioning of CF tips. Then two CFs were attached to both cell ends from the top surface of the cell to apply axial stretch to the cell (Figure 1a). Both CFs received the same control command to achieve identical CF position control for stretching the cell end. All experiments were carried out at a stimulation frequency of 1 Hz at room temperature.

The measurement of mechanical activity of isolated cardiomyocyte was performed by the laser confocal scanning microscopy system LSM 710 (Carl Zeiss, Germany) and the software developed for fast acquisition of cell image (ZEN 2010, Calr Zeiss, Germany). Prior to the measurement, a narrow region of scanning was selected on cell image, assuming it contained both CFs (Figure 2a). The intensity profile within the region was then acquired continuously every 1-3 msec. The distance between the CFs was calculated in this profile from the two areas with lowest intensity while striation pattern between the areas was further analyzed to reconstruct mean sarcomere length on the basis of Discrete Fourier Transform (Figure 2b).

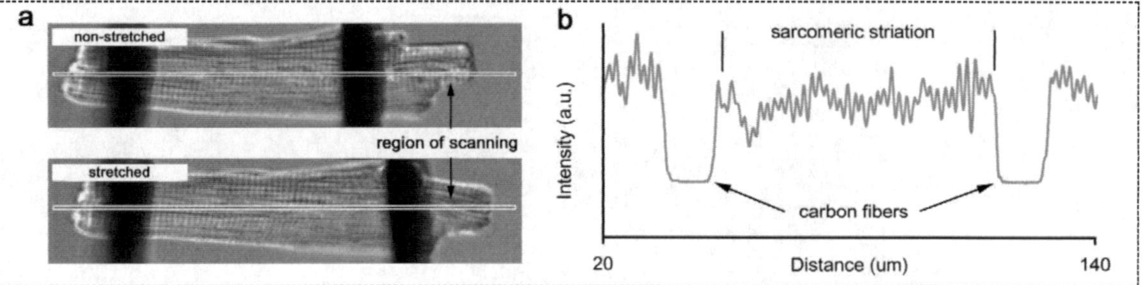

Figure 2. The technique of carbon fibers (CFs) applied for stretching a single cardiomyocyte. (a) Two images of a cell (non-stretched and stretched by CFs). The long and narrow region indicates the area where image acquisition is performed. (b) The example of intensity profile obtained from a region of scanning. The two areas with lowest intensity correspond to the positions of CFs, striation pattern between the areas is used to retrieve mean sarcomere length.

It should be noted that the effectiveness of CF technique strongly depends on the quality of CFs adhesion to cell surface. Figure 3 shows the schematic examples of adequate and inappropriate adhesion of CFs to cell surface. In some cases, adhesion of CFs to the cell surface was sufficiently good to apply adequate stretch to cardiomyocytes and their sarcomeres, i.e. CF sliding to the cell ends is not observed (Figure 3a). In other cases, CF adhesion was not appropriate, so one could see gradual sliding of CFs to the edges of the cell (Figure 3b) when stretch command by PZTs was applied. As provided by the example in Figure 3, the measured cell length, if evaluated as a distance between two CFs, increases with the stretch command in the same extent for both cases. However, in the case of inappropriate CFs adhesion the extent of sarcomere stretch is obviously deficient compared to the measured cell length. In this case, the actual cell length, if obtained directly from the measured length, is overestimated, while actual sarcomere length is much lower than it can be expected from the measured cell length.

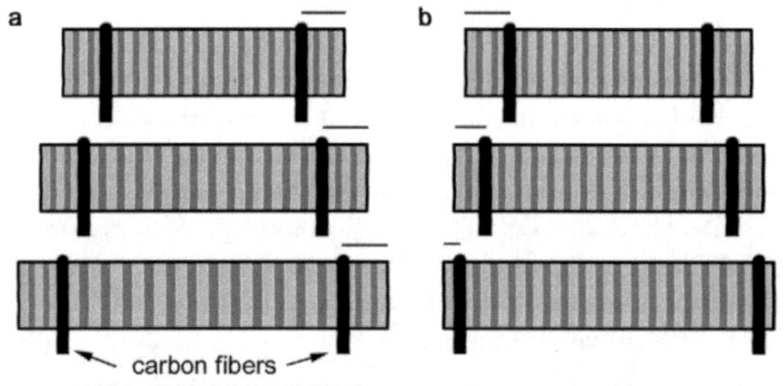

Figure 3. The schematic examples of adequate (a) and inappropriate (b) adhesion of CFs to the cell surface when stretch was applied. Horizontal lines located at the top-right end of the cell indicate the amount of sliding (the smaller line the higher sliding).

3. Results

Figure 4 shows representative contractions of isolated ventricular cardiomyocyte at gradually increased end-diastolic cell lengths (upon stretch by CFs). The upper trace, i.e. cell length, was obtained directly from the positions of left and right CFs (Figure 4a). The bottom trace, sarcomere length, was retrieved from striation pattern between the two CFs (Figure 4b). As one could see in Figure 4, the extent of cell stretch is somewhat higher than that of sarcomere stretch.

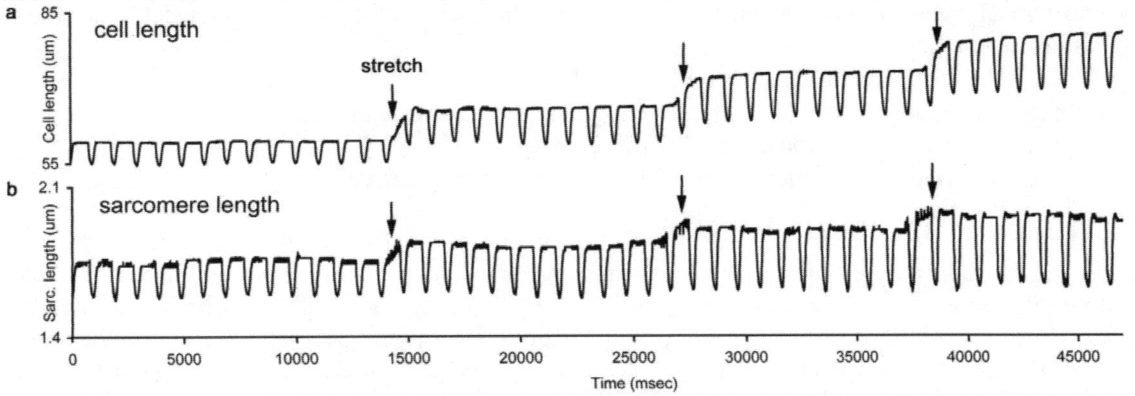

Figure 4. The representative contractions of isolated ventricular cardiomyocyte recorded at gradually increased end-diastolic cell lengths. (a) Cell length trace obtained directly as the distance between left and right carbon fibers. (b) Sarcomere length trace. Arrows show the moments of stretch applied.

In order to estimate the possible discrepancy between the amount of cell stretch and SL stretch, we analyzed the "end-diastolic cell length - end-diastolic SL" relationship in ventricular cardiomyocytes isolated from the rat heart (n = 221) and the guinea pig heart (n = 67). For most cells the dependence had a highly linear shape, so each of them was approximated by a linear function $y = kx + b$.

Figure 5 shows representative "end-diastolic cell length - end-diastolic SL" relationships, expressed as a percent of the initial (non-stretched) cell length / initial SL. A theoretical curve has a linear coefficient $k = 1$, so the stretch amount of the cell strictly corresponds to the stretch amount of its sarcomeres (Figure 5, curve #1). For some cells we observed that an increase in the cell length was always greater than in its sarcomeres but by a constant factor at all displacements ($k < 1$, Figure 5, curve #2); in this case the curve remains highly linear. For some cells we observed a gradual non-linear deviation from the theoretical curve ($k < 1$, Figure 5, curve #3); in this case the amount of CFs sliding increased non-proportionally with the extent of stretch command and the curve has highly non-linear behavior.

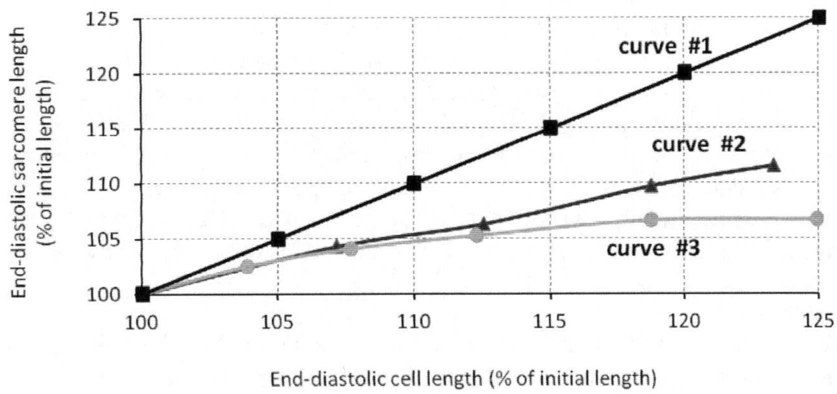

Figure 5. Representative "end-diastolic cell length - end-diastolic SL" curves. Curve #1 represents a theoretical curve (with a linear coefficient k=1), curve #2 displays a proportional deviation from the theoretical curve and curve #3 displays a gradual non-linear deviation from the theoretical curve; curves #2 and #3 have a linear coefficient k<1.

Prior to the selection of cells for further analysis, the following criteria were applied for the linear regression approach:

- the number of stretch steps for each cell was ≥ 3;
- the maximum percent of cell stretch relative to its initial length was ≥ 10%, so this allowed us to assess the CF adhesion at large displacements;
- the coefficient of determination for linear regression for each cell was ≥ 0.9.

Accordingly to the criteria, further analysis was carried out for n = 181 rat cardiomyocytes and n = 42 guinea pig cardiomyocytes. For each cell, a linear coefficient of "end-diastolic cell length - end-diastolic SL" relationship was calculated and used as a quality index of CF adhesion to the cell. Further, the distributions were checked for normality. We found that mean value of the linear coefficient was ~0.3 for the rat cardiomyocytes and ~0.7 for the guinea pig cells (Figure 6) which means that CFs sliding was observed almost in all experiments.

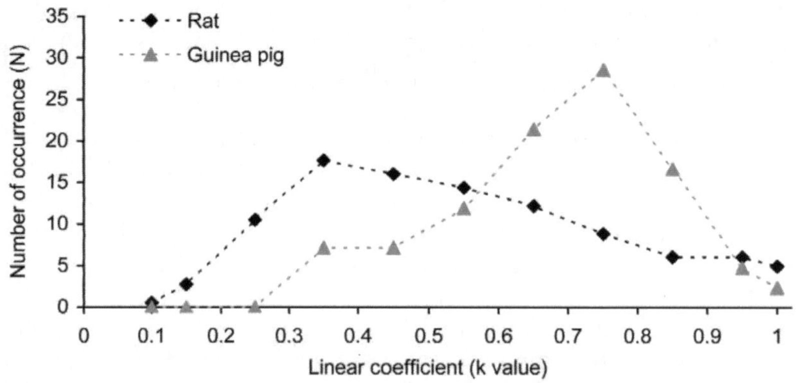

Figure 6. The normal distribution of linear coefficient values (k) for "end-diastolic cell length - end-diastolic sarcomere length" relationships obtained for rat and guinea pig ventricular cardiomyocytes.

4. Conclusions

CF approach remains a unique and sophisticated technique to study single cardiomyocyte contractility when cells exposed to mechanical load (applied viastretch in our study). A limitation of this technique is the absence of overall control of CF sliding relative to the cell surface, which causes a deviation of the cell length changes against its sarcomeres. In this study, we presented a method to assess CF sliding using a CF adhesion quality index, which can be defined as a linear coefficient from "end-diastolic cell length - end-diastolic SL" relationship. The next step for the development of the method should be an assessment of the allowable range of the index values in order to analyze cell contractility eliminating the cells with huge CF sliding from the data samples but to do not reduce to zero the success rate and to get experimental data using such limited CF technique in itself.

5. Acknowledgements

This study was supported by Russian Basic Research Foundation (#16-04-00085 to OL, #16-31-60015 to AK) and RF Government Resolution #211 of March 16, 2013 to TM.

6. References

[1] Friedrich O, Schneidereit D, Nikolaev YA, Nikolova-Krstevski V, Schürmann S, Wirth-Hücking A, et al. Adding dimension to cellular mechanotransduction: Advances in biomedical engineering of multiaxial cell-stretch systems and their application to cardiovascular biomechanics and mechano-signaling. Progress in Biophysics and Molecular Biology. 2017.

IOP Conf. Series: Materials Science and Engineering **350** (2018) 012011 doi:10.1088/1757-899X/350/1/012011

[2] Clemo HF, Stambler BS, Baumgarten CM. Swelling-activated chloride current is persistently activated in ventricular myocytes from dogs with tachycardia-induced congestive heart failure. Circulation Research. 1999;84(2):157-65.

[3] Dyachenko V, Husse B, Rueckschloss U, Isenberg G. Mechanical deformation of ventricular myocytes modulates both TRPC6 and Kir2. 3 channels. Cell calcium. 2009;45(1):38-54.

[4] Jian Z, Han H, Zhang T, Puglisi J, Izu LT, Shaw JA, et al. Mechanochemotransduction during cardiomyocyte contraction is mediated by localized nitric oxide signaling. Science signaling. 2014;7(317):ra27.

[5] Iribe G, Helmes M, Kohl P. Force-length relations in isolated intact cardiomyocytes subjected to dynamic changes in mechanical load. American Journal of Physiology-Heart and Circulatory Physiology. 2007;292(3):H1487-H97.

[6] Iribe G, Kaneko T, Yamaguchi Y, Naruse K. Load dependency in force–length relations in isolated single cardiomyocytes. Progress in biophysics and molecular biology. 2014;115(2):103-14.

[7] Nishimura S, Yasuda S-i, Katoh M, Yamada KP, Yamashita H, Saeki Y, et al. Single cell mechanics of rat cardiomyocytes under isometric, unloaded, and physiologically loaded conditions. American Journal of Physiology-Heart and Circulatory Physiology. 2004;287(1):H196-H202.

[8] Peyronnet R, Bollensdorff C, Capel R. A, Rog-Zielinska E. A, Woods C. E, Charo D. N, Lookin O, et al. Load-dependent effects of apelin on murine cardiomyocytes. Progress in Biophysics and Molecular Biology. 2017.

[9] Le Guennec J-Y, Peineau N, Argibay J, Mongo K, Garnier D. A new method of attachment of isolated mammalian ventricular myocytes for tension recording: length dependence of passive and active tension. Journal of molecular and cellular cardiology. 1990;22(10):1083-93.

[10] Yasuda S-I, Sugiura S, Kobayakawa N, Fujita H, Yamashita H, Katoh K, et al. A novel method to study contraction characteristics of a single cardiac myocyte using carbon fibers. American Journal of Physiology-Heart and Circulatory Physiology. 2001;281(3):H1442-H6.

Comparative Study of the Physical, Topographical and Biological Properties of Electrospinning PCL, PLLA, their Blend and Copolymer Scaffolds

E Bolbasov[1], S Goreninskii[1], S Tverdokhlebov[1], A Mishanin[2], A Viknianshchuk[3], D Bezuidenhout[4] and A Golovkin[2]

1 National Research Tomsk Polytechnic University, 30 Lenin av., Tomsk, Russian Federation 634050
2 Almazov National Medical Research Centre, 2 Akkuratova str., St. Petersburg, Russian Federation 197341
3 Saint Petersburg University, 13B Universitetskaya Emb., St. Petersburg, Russian Federation 199034
4 University of Cape Town, 203 Cape Heart Centre, Anzio Road, Observatory, Cape Town, South Africa 7925

Abstract. Biodegradable polymers (blends, copolymers) could be the ideal materials for manufacturing of scaffolds for small diameter vascular graft. Such material characteristics as mechanical properties, chemical structure, nano- and micro topography, surface charge, porosity, wettability etc. are becoming the most important aspects for effectiveness of prosthesis biofunctionalization because of their great impact on cell adhesion, spreading, cell proliferation, differentiation and cell function. The aim of the study is to compare physical, topographical and biological properties of polycaprolactone (PCL), poly-L-lactic acid (PLLA), polycaprolactone + poly-L-lactic acid blend (PCL PLLA), L-lactide/Caprolactone copolymer (PLC7015) scaffolds fabricated with the same fiber thickness using electrospun technology. PCL PLLA scaffolds had the highest average pore area ($p<0.01$) and the lowest strength ($p<0.01$). PLC7015 scaffolds had the significantly lower average pore area ($p=0.03$) but the highest elastic deformation ($p<0.01$). Biological testing with MMSC (multipotent mesenchyme stem cells) demonstrated that after 72 hours of co-cultivation only on PCL and PLLA scaffolds cells entered to the active phase of adhesion process. We propose that physical and topographical properties of PCL, PLLA, their blend and copolymer are of a great dependence of chemical structure but could be changed during the manufacturing process that will lead to changes in biological properties.

1. Introduction

Autologous veins and arteries remain the only option for coronary artery bypass surgery and the vessels of choice for peripheral small diameter applications. When suitable natural vessels are unavailable for peripheral grafting, durable synthetic prostheses (ePTFE and Dacron) with poor long-term patencies are the only current alternative [1]. There is thus a large drive towards the development of tissue engineering and regenerative medicine (TERM) approaches to vascular grafts in order to not only provide suitable short-term solutions, but also to generate living neo-arteries that function as well as native vessels [2].

We propose that biodegradable polymers (blends, copolymers) could be the ideal materials for manufacturing of scaffolds for such small diameter vascular graft, as they have some advantages that could be useful for these purposes. For instance polymers have more altitudinal synthetic flexibility with a variety of functional groups; multitudinous functional polymers show excellent mechanical and chemical robustness; polymers are easily patterned with state-of- the-art nano- and microtechnologies; many polymers are nontoxic and scalable, which are critical for real applications [3].

Thanks to the advantages of electrospun nanofibrous scaffolds, such as potential for mimicking of the extracellular matrix, nontoxicity, biocompatibility, biodegradability, permeability and low cost, PCL (poly(ε-caprolactone)) and PLLA (poly-L-lactic acid) are very promising for vessel graft development [4][5]. PCL contains significant number of alkyl units between the ester links, has more flexible polymer chain, low crystallization temperature, rubber-like behavior and a very long biodegradation period. PLLA contains more complex ester units in the chain, so it is more susceptible to degradation by hydrolysis and shows faster biodegradability but its applications are limited because of its lack of toughness. PLLA is rigid in a room temperature and at body temperature. A blend, or copolymer of PCL and PLLA, could therefore theoretically combine the properties of both polymers and provide resistant but ductile mechanical behavior, as well as prolonged biodegradability [6].

The cell-polymer interactions become the key questions of implanted prosthesis functioning and its biocompatibility. Such material characteristics as mechanical properties, chemical structure, nano- and micro topography, surface charge, porosity, wettability etc. are becoming the most important aspects for effectiveness of prosthesis biofunctionalization because of their great impact on cell adhesion, spreading, cell proliferation, differentiation and cell function. In this case choose of polymer for the scaffold becomes very important.

It is known that the set of physical, chemical and biological properties of nanofibrous scaffolds developed by electrospinning is largely determined by the fiber thickness. There are some isolated investigations aimed to find the fiber thickness for optimal physical, chemical and biological properties of PCL, PLLA, their copolymers and blends for vessel grafts [7]. Meanwhile there are no comparative studies between samples from different chemical groups that makes difficult to choose the right polymer or copolymer or blend for vessel graft development.

The aim of the study is to compare physical, topographical and biological properties of polycaprolactone, poly-L-lactic acid, polycaprolactone + poly-L-lactic acid blend, L-lactide/Caprolactone copolymer scaffolds fabricated with the same fiber thickness using electrospun technology.

2. Material and methods

Poly(ε-caprolactone) (molecular weight 70 000 g/mol) (PURASORB® PC 12, PURAC) (PCL), Poly-L-lactic acid (molecular weight 650 000 g/mol) (PURASORB® PL 38, PURAC) (PLLA) and L-lactide/Caprolactone copolymer (molecular weight 200 000 g/mol) (PURASORB® PLC 7015, PURAC) (PLC7015) were obtained from Corbion, and a Poly-L-lactic acid + Polycaprolactone blend (PCL PLLA) in a 70/30 molar ratio prepared by admixture. Solutions of the polymers in hexafluoro-2-propanol (≥99%, Sigma-Aldrich) were prepared in a hermetic glass reactor at a room temperature with a constant vortexing. Final concentrations of solutions were 7wt% for PCL, 4wt% for PLLA, 4wt% for PCL PLLA, 4wt% for PLC7015. Scaffolds were fabricated using electrospinning (NANON-1, MECC Co., Japan) at 30V potential difference using a 10x200mm (diameter x length) collector rotated at 50 rpm, 140 mm needle (22G) to target distance . Solution pump rates were optimized for each solution to achieve similar fibre thicknesses: 3 ml/hour for PCL; 5,3 ml/hour for PLLA; 5 ml/hour for PCL PLLA; 5 ml/hour for PLC7015.

Micrographs of graft morphology were obtained by means of scanning electron microscopy (JCM-6000Plus, Jeol) after gold sputtering (SmartCoater, Jeol). Resulting SEM-images were analyzed by using Image J software (National Institutes of Health, USA): fibre thickness by direct measurement, coherence using the Orientation J plugin (Biomedical Engineering group, EPFL), and pore size using a bespoke macro (Biomaterials Group, UCT) (Figure 1).

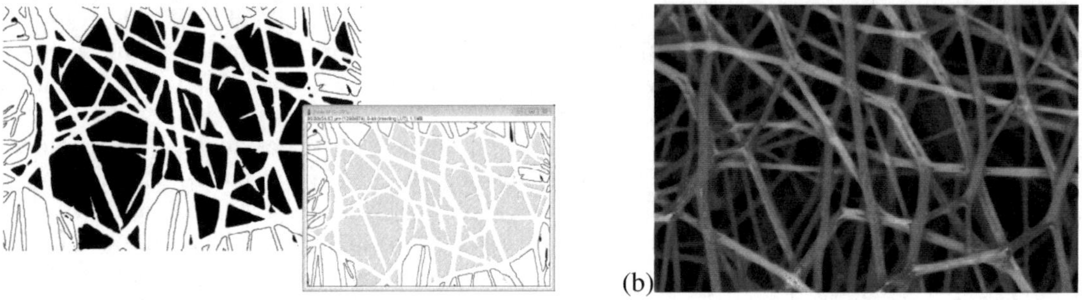

(a) (b)

Figure 1. Example of (a) pore size quantification after elimination of edge effects and (b) false colour map assigned to fibre directions in the determination of the coherency of the scaffold fibres.

Water contact angle of the grafts surface was measured by sessile drop method (EasyDrop, Kruss). Mechanical properties of the grafts were studied under uniaxial extension using Instron 3369 installation with 50N loading cell (model 2519-102, Instron) at a loading velocity of 10 mm/min in a longitudinal direction.

For biological testing, fat-derived MMSC (multipotent mesenchyme stem cells) were collected from healthy donors and immunophenotyped with flow cytometer GuavaEasyCyte8 (Millipore, USA) using CD19, CD34, CD45, CD73, CD90 and CD105 monoclonal antibodies (BD, USA). To analyze the cell adhesion and attachment, focal adhesion vinculin was studied using fluorescence microscopy. Samples with cells after 72 hours of incubation discarded from medium, the cells washed with PBS and fixed with 4% paraformaldehyde for 20 min. Cells were permeabilised using Triton X-100 and rinsed with PBS, blocked with 10% goat serum in PBS for 30 min at room temperature, and incubated with anti-vinculin antibody (Thermo Scientific) at 1:200 dilution for 2 hours. After 3 PBS washes cells were incubated with Alexa Fluor 488 goat anti-mouse IgG (H+L) (Invitrigen) at 1:200 dilution for 1 h at room temperature in the dark. The cells were washed three times in PBS (5 min each) and stained with 4′,6 diamidino-2-phenylindole (DAPI) for nuclear visualization. Finally, the cells were washed and viewed using Carl Zeiss Axio Observer fluorescence microscope. Images were collected and processed with Zen Software. Cells were recalculated as their number per mm2.

To verify cells viability after co-cultivation with samples they were trypsinised from the material and stained with Annexin V FITC (Biolegend) and Propidium Iodide (Sigma Aldrich) for 20 minutes in the dark. Flow cytometry was performed on GuavaEasyCyte8 with detection of double positive (late apoptosis/necrosis), double negative (living) and annexin V positive (apoptosis) cells in relative count in percent. All biological tests were run at least in triplicates.

Statistical analysis was performed using Statistica 7.0 software (StatSoft, USA). Data is presented as Mean±SE (standard error of mean). The significance of difference was calculated using the Mann-Whitney U-test. Differences between groups were stated as significant in $p<0.05$.

3. Results

All developed nanofibrous scaffolds were formed by randomly entangled fibers of a cylindrical shape with a diameter of 1.2-1.3 μm and with normality in size distribution. All nanofibrous scaffolds were hydrophobic (Figure 1, Table 1) due to the low surface energy of polymers and high surface porosity [8].

The slightly lower values of contact angle for PLLA scaffolds are probably due to the greater number of functional groups on the fibers surface caused by chemical nature of the polymer material.

Scanning electron microscopy images of prepared nanofibrous scaffolds are presented in the Figure 2.

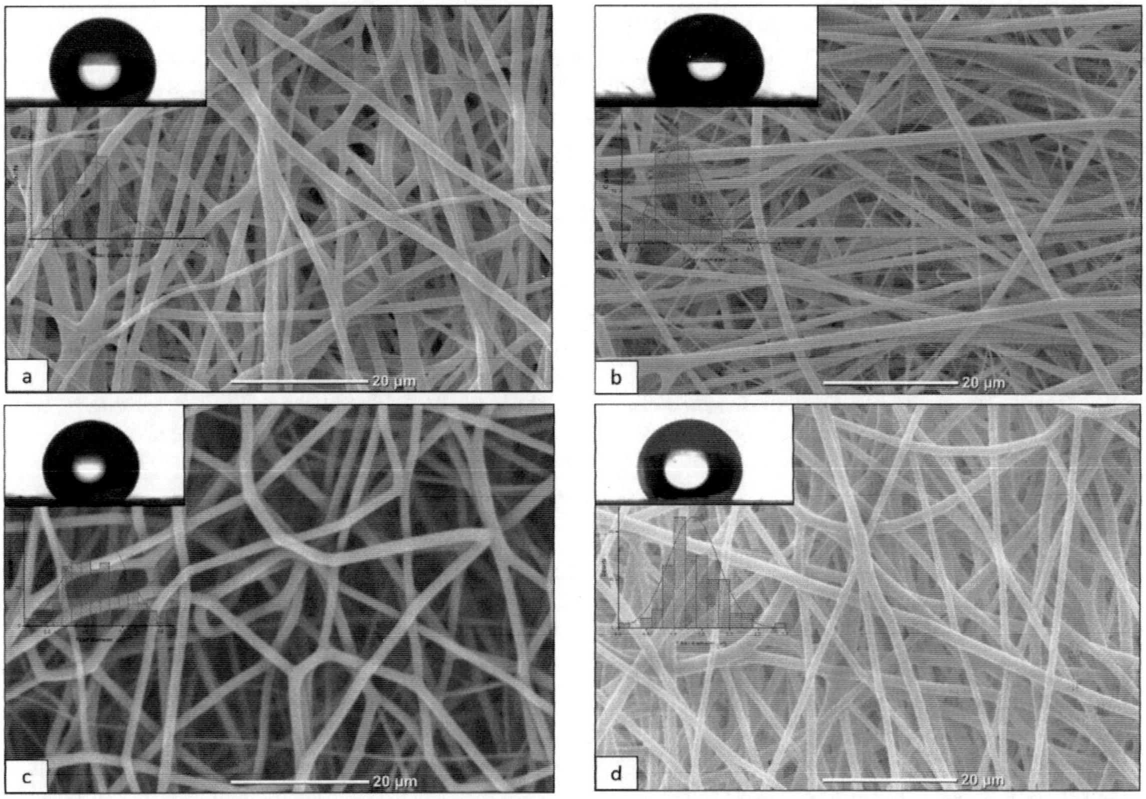

Figure 2. Scanning electron microscopy and wettability measurement of prepared nanofibrous scaffolds. A-PCL, B-PLLA, C -PCL PLLA, D-PLC7015.

The scaffolds also had similar random orientation of the fibres (average coherence = 0.13±0.06) and similar pore sizes (average 11.2±5.0μm^2). See Table 1.

The highest strength values and the smallest value of the relative elongation at uniaxial stretching were in PLLA nanofibrous scaffolds (Table 1). It is caused by completely non-crystalline but highly oriented polymeric chains in fibers forming scaffolds [9]. PCL and PLC7015 nanofibrous scaffolds had the intermediate strength values due to the semicrystalline nature of these polymers at room temperature. Meanwhile having a lower crystallinity degree compared to PCL, PLC7015 has the higher number of passing molecules between the crystal regions, which provides the highest values of relative elongation [10][11]. The lowest strength was in PCL PLLA blend due to thermodynamic incompatibility of these polymers at the molecular level [12]. On the one hand this forms high fibers defectiveness in molecular level and on the other hand PLLA hampers the crystallization of PCL at room temperature forming a significant amount of macromolecules in the amorphous state and providing the high values of relative elongation of this type of scaffolds.

Table 1. Fiber thickness, wettability, topographical and mechanical properties of prepared nanofibrous scaffolds, Mean±SE

Material	Average fiber diameter, μm	Average pore area, μm^2	Contact wetting angle, degree	Strength, Mpa	Elastic deformation, %
PCL	1.3 ± 0.5	9.5±0.6	123 ± 2	8.2 ± 0.7	112 ± 10
PLLA	1.2 ± 0.4	8.9±0.5	112 ± 2[a]	13.2 ± 1.0[a]	95 ± 10[a]
PCL PLLA	1.2 ± 0.4	18.7±1.2[a]	128 ± 2	4.7± 0.3[a]	265 ± 44[a]
PLC7015	1.3 ± 0.5	7.8±0.5[b]	123 ± 2	9.1 ± 1.9	560 ± 157[a]

[a] $p<0.01$ comparing with PCL

[b] $p=0.03$ comparing with PCL

Materials with the same fibers thickness had different adhesion properties (Table 2, Figure 3). PLC 7015 had the lowest level of adhered MMSC and all other materials had the results as PCL. Meanwhile cell viability – parameter that basically related with chemical properties and direct toxicity – showed that PLC7015 and PCL PLLA blend had the equal percent of living cells on their surface and it was higher than those on PCL (p=0.05).

Table 2. Cell viability and cell adhesion on prepared nanofibrous scaffolds, Mean±SE.

Polymer	Cell viability, %			Cell adhesion, cells/mm^2
	Living cells	Late apoptosis/ necrosis	Apoptosis	
PCL	81.66±1.17	7.18±0.63	10.05±0.48	129±10
PLLA	77.24±0.70 [a]	11.87±1.24[a]	11.53±0.44[a]	143±10
PCL PLLA	88.36±0.32[a]	3.90±0.15[a]	6.42±0.37[a]	126±15
PLC7015	88.50±1.63[a]	4.79±0.44[a]	4.76±1.15[a]	64±3[b]

[a] p=0.05 comparing with PCL
[b] p<0.01 comparing with PCL

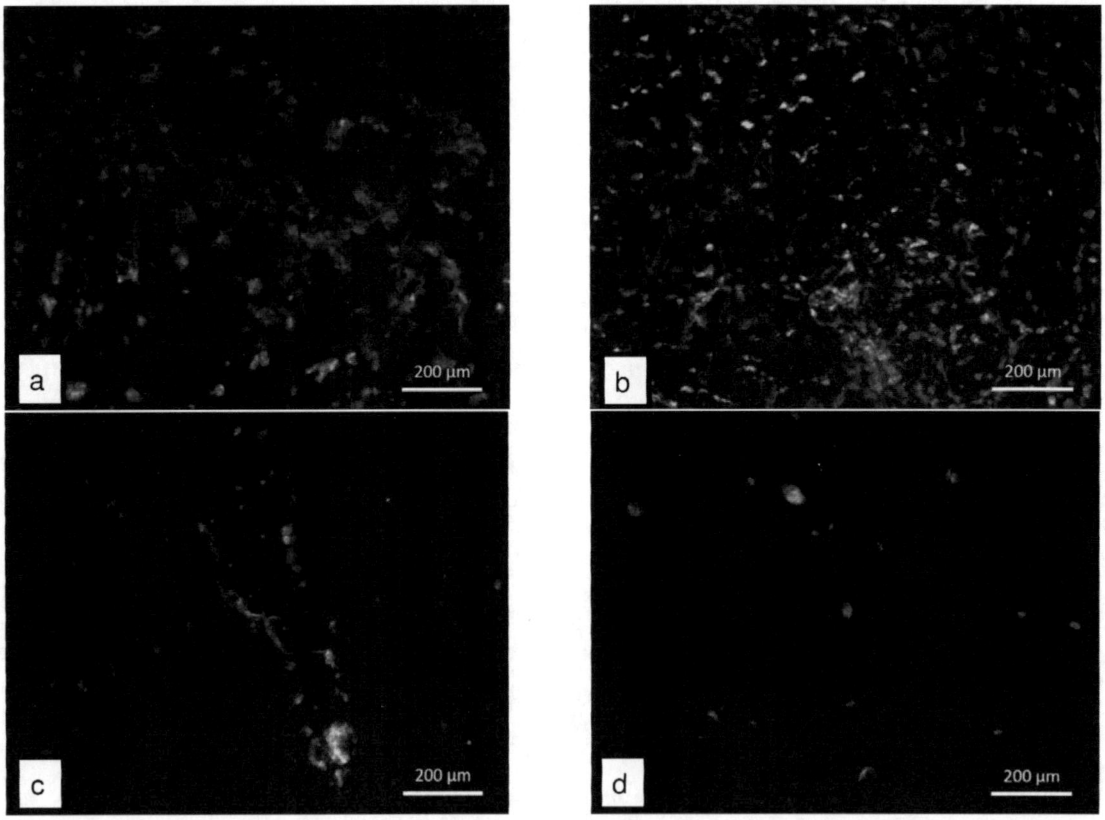

Figure 3. Adhered MMSC stained with vinculin (green) and DAPI (blue) on prepared nanofibrous scaffolds. A-PCL, B-PLLA, C -PCL PLLA, D-PLC7015.

Cells on the fibers of PCL PLLA and PLC7015 were not well spread and there were no focal adhesions (compaired to PCL and PLLA scaffolds). Taken together with adhesive cell level these data indicate that MMSC had a good first phase when the complex physicochemical interactions including hydrophobic, coulombic, and van der Waals forces between the cell membrane and the polymer surface leads to primary adhesion [3]. All interactions during this process are passive adhesion. On the next active phase arising from the integrin binding, cells start to spread and become flattened. After 72 hours of co-cultivation of MMSC with polymers surface this is observed only on the PCL and PLLA scaffolds. The third phase, when the cytoskeleton is organized to focal adhesion between cells and material was not achieved in the materials tested in this study.

4. Discussions

Different polymers, copolymers and blends could have different chemical structure which could have the impact on their properties. Besides it is proposed that nano- and micro topography, surface charge, porosity etc. are the main characteristics that directly influence on biocompatibility of the material and they are basically depended from the used manufacturing technique [3].

Synthetic biodegradable polymers such as PCL, PLLA, their blend and copolymer are the most promising materials for vascular tissue engineering and they are widely tested for different applications. Meanwhile the comparative study of electrospun materials with the close or same morphology is needed for correct chooses of the material. Besides because of high cost of copolymers comparing with blends it could be useful according to economic considerations [6].

Previously it was shown that hemolysis, coagulation pathways and thromogenicity of PCL and PLC7015 electrospun scaffolds depends on fibrous morphology and chemical composition of the scaffolds [7]. That means that adhesion properties of polyesters could be of a great addiction on material morphology in particular on fiber thickness and porosity. Also these parameters could lead to change in physical and chemical properties of the scaffolds.

Additionally, along with the material choice, it is critical to use optimal scaffold development techniques even if electrospinning is chosen as a basic technology, as the purpose is to fabricate scaffolds with polymer surfaces that mimic natural ECM (extracellular matrix) with different chemical, mechanical and topographical cues which facilitate interactions between cells and materials to control and influence cell behavior.

The impact of the chemical structure of polymers on biocompatibility is often performed using solid membranes, without taking the morphology of scaffold into account. However, by choosing the stock polymer first and comparing with others that are developed with the same ECM mimicing topography, is preferred.

Our results show that even the same topography of different polymer electrospun scaffolds could result in different biocompatibility properties. The chemical structure of the used polymers is privileged but it also means that trying to create mimic ECM one should find the balance between the polymer and optimal topography. It is also well known that surface topography influences cell adhesion, migration and proliferation. As the alignment of electrospun fibres can influence the interaction with cells, the scaffolds described in the current paper were specifically produced to have similar high randomness, as evidenced by the low coherencies of the four scaffold types. Similarly, it is evident that porosity, and specifically pore size, could influence cell adhesion and penetration, and equal care was taken to exclude this parameter as a variable.

The results show that the strength of prepared copolymer and blend scaffolds was differing than those in pure composed polymers. Copolymer PLC7015 demonstrated the same strength as PCL but the lower then PLLA. Meanwhile it has the highest level of elastic deformation. Properties of PCL PLLA blend were significantly differ then those in copolymer. The lower strength was combined with higher than in PCL scaffold elastic deformation. Actually strength of the vessel graft is a very important for long functioning in a blood flow without complications related with the hemorrhage or bleeding. Meanwhile due to elastic deformation the pulsatile wave is extinguished and it could prevent critical deformations of the graft but besides the high level of deformation could lead to aneurism of the prosthesis and to its thrombosis or disruption.

Adhesive properties and cytotoxicity of scaffolds are the main characteristics of biocompatibility for their long functioning in the blood flow. Testing these properties in vitro show the possibilities of cellularization or endothelialization of the grafts which are the first steps for biofunctionalization and future biodegradation.

Our results show that due to the used manufacturing technology and created graft topography PLC7015 copolymer have the lowest adhesive properties but adequate cytotoxicity. All other polymers and blend were rather equal in MMSC adhesion on their surfaces. We suppose such low adhesion is due to the lower average rope area. And our preliminary results (data not shown) lead us to proposal that PLC7015 could have excellent adhesive properties which will be much more attractive then PCL or PLA itself but in the case of usage of manufacturing technique that will allow high porosity, fiber thickness, fiber heterogeneity etc.

5. Conclusion

Physical, topographical and biological properties of polycaprolactone, poly-L-lactic acid, polycaprolactone + poly-L-lactic acid blend, L-lactide/Caprolactone copolymer scaffolds fabricated with the same fiber thickness using electrospun technology could be significantly different and the chemical properties of the polymers (copolymers, blends) become the leading factor of the changed characteristics that lead to different biocompatability. Physical and topographical properties could be changed during the manufacturing process that will lead to changes in biological properties. Physical, topographical and biological properties of PCL and PLLA copolymer and blend could greatly vary from the stock polymers and these characteristics could be managed by using different electrospinning protocols.

6. References

[1] Zilla P, Bezuidenhout D, Human P. Prosthetic vascular grafts: Wrong models, wrong questions and no healing. Biomaterials. 2007;28:5009–27.
[2] Khademhosseini A, Langer R. A decade of progress in tissue engineering. Nat. Protoc. 2016;11:1775–81.
[3] Chen L, Yan C, Zheng Z. Functional polymer surfaces for controlling cell behaviors. Mater. Today. Elsevier Ltd; 2017;00. https://doi.org/10.1016/j.mattod.2017.07.002
[4] Mugnai D, Tille JC, Mrówczyński W et al. Experimental noninferiority trial of synthetic small-caliber biodegradable versus stable vascular grafts. J. Thorac. Cardiovasc. Surg. 2013;146
[5] Sevost'ianova VV, Elgudin IA, Glushkova TV, Wnek G, Liubysheva T, Emancipator S, Kudriavtseva IuA, Borisov VV, Golovkin AS, Barbarash LS Use of polycaprolactone grafts for small-diameter blood vessels. Angiol. Sosud. Khirurgiia Angiol. Vasc Surg. 2015;21:44–53
[6] Lebourg, M., Anton Suay J, Ribelles Gomez JL. Porous membranes of PLLA-PCL blend for tissue engineering applications. Eur. Polym. J. 2008;44:2207–18
[7] Horakova J, Mikes P, Saman A et al. Comprehensive assessment of electrospun scaffolds hemocompatibility. Mater. Sci. Eng. C. 2017 doi: 10.1016/ j.msec.2017.05.011
[8] Cassie ABD, Baxter S. Wettability of porous surfaces. Trans. Faraday Soc. 1944;40:546
[9] Zong X, Kim K, Fang D et al. Structure and process relationship of electrospun bioabsorbable nanofiber membranes. Polymer (Guildf). 2002;43:4403–12
[10] Mo X., Xu C., Kotaki M et al. Electrospun P(LLA-CL) nanofiber: a biomimetic extracellular matrix for smooth muscle cell and endothelial cell proliferation. Biomaterials. 2004;25:1883–90
[11] Wang X, Zhao H, Turng L-S et al. Crystalline Morphology of Electrospun Poly(ε-caprolactone) (PCL) Nanofibers. Ind. Eng. Chem. Res. 2013;52:4939–49
[12] Dell'Erba R, Groeninckx G, Maglio G et al. Immiscible polymer blends of semicrystalline biocompatible components: thermal properties and phase morphology analysis of PLLA/PCL blends. Polymer (Guildf). 2001;42:7831–40.

Acknowledgments

The authors wish to acknowledge the assistance of Wian van den Bergh for the pore size macro and determination.
This research was funded by Russian Science Foundation (project № 16-13-10239) and performed in Tomsk Polytechnic University and Almazov National Medical Research Centre.

IOP Conf. Series: Materials Science and Engineering **350** (2018) 012013 doi:10.1088/1757-899X/350/1/012013

Hydrodynamic Assists Magnetophoreses Rare Cancer cells Separation in Microchannel Simulation and Experimental Verifications

O Saeed[1,2], L Duru[1] and D Yulin[1]

1 School of Life Science, Department of Biomedical Engineering, Beijing Institute of Technology, Beijing 100081, China
2 Faculty of science and technology, Department of Biomedical Physics, AL-Neelain University, Khartoum 1121, Sudan
E-mail: omerphy2007@bit.edu.cn

Abstract. A proposed microfluidic design has been fabricated and simulated using COMSOL Multiphysics software, based on two physical models included in this design. The device's ability to create a narrow stream of the core sample by controlling the sheath flow rates Qs_1 and Qs_2 in both peripheral channels was investigated. The main target of this paper is to study the possibility of combing the hydrodynamic and magnetic techniques, in order to achieve a high rate of cancer cells separation from a cell mixture and/or buffer sample. The study has been conducted in two stages, firstly, the effects of the sheath flow rates (Qs1 and Qs2) on the sample stream focusing were studied, to find the proposed device effectiveness optimal conditions and its capability in cell focusing, and then the magnetic mechanism has been utilized to finalize the pre-labelled cells separation process.

1. Introduction

The use of single and pure cells in biological field represents a vital required demand for therapy, drug discovery and disease control and monitoring. As a result, cell separation, particularly of circulating tumour cells (CTCs) isolation and separation from a blood sample has gained high interest in the field of biology and biotechnology.

Due to the fact that, the abundance of the CTCs in the blood can be as low as 1 in 10^9 erythrocytes and 1 in 10^7 leukocytes, and they somehow share characteristics and properties with leukocyte cells, their separation represents a challenges for researchers. The design of microfluidic platforms that have a high and precise ability in CTCs separation has been proven difficult. However, erythrocytes have their own unique physical, chemical and biological properties, which facilitate their separation and makes it easy to remove them from the blood[1][2].

Recently, several methods and mechanisms have been established for CTCs separation, they can generally be classified into two broadly approaches, firstly, biochemical methods, which were mainly achieved by merging the cell surface antigens express by the target cells, and intracellular markers fluorescent tagging to identify the CTCs. The CTCs identify as positive for cytokeratins (CK) fluorescent staining, while leukocytes were defined by staining for CD45 (receptor-linked protein tyrosine phosphatase).

The separation based on the second approach (biophysical methods) is totally based on the existence of differences of physical properties between CTCs and leukocytes cells e.g. size [3][4][5]. Deterministic lateral displacement (DLD) devices are also a part of this approach, which has been

IOP Conf. Series: Materials Science and Engineering **350** (2018) 012013 doi:10.1088/1757-899X/350/1/012013

utilized for separation based on cells and particles sizes [6][7]. Cell sorting techniques, as reviewed above and elsewhere [8][9], are limited in yield and purity, furthermore, the operation of these methods relies highly on the expertise of skilled personnel. Different fabricated microfluidic platforms have shown their high capability in the detection and separation of targeted cells from a mixture, which is highly demanded in biological and clinical research. Due to low costs, small sample volumes, fast processing times, large surface-to volume and the ability to efficiently isolating single cell, and thanks to features available in microfluidic platforms[10],different methods based on microfluidic cell separation were established including dielectrophoresis [11] , acoustophoresis[12] and hydrodynamics [13].
There are many microfluidic platforms that have been studied and utilized for CTCs separation based on biochemical methods [14][15]. Anti-epithelial cell adhesion molecule antibody (anti-EpCAM antibody), has been commonly used in studies for CTCs identification as U.S. FDA (Food and Drug Administration) approved Veridex Cellsearch® system (USA), considered as an example in utilizing this approach for clinical uses[16].
As shown in our previously published review paper[17], the different microfluidic methods that have been fabricated to separate cells including CTCs, suffers some limitations. Until now a few work has been done to overcome these limitation, by utilizing a microfluidic platforms that combine and merge two or more technique into one microfluidic platform. Here we are trying to overcome the one technique limitations, by utilizing a microfluidic device used to separate CTCs spiked in blood and /or buffer sample. Firstly, a hydrodynamic module used to concentrate and align the CTCs contained the sample in a narrow stream by controlling the two sheath flow rates Qs in both peripheral channels. Secondly, exerting a magnetic force on the pre-labelled magnetic beads CTCs, generated by a permanent magnet located at specific distance from the main channel wall, enable the separation of pre-labelled CTCs from cell mixture and/or buffer sample. The study has been conducted in two stages, firstly the effect of both sample flow rate $\mathbf{Q_i}$ and the two sheath flow rates $\mathbf{Q_{s1}}$ and $\mathbf{Q_{s2}}$ on sample stream focusing were studied, to find the proposed device effectiveness optimal conditions and its capability, then the magnetic mechanism was utilized to finalize the separation proses of the pre-labelled cells with magnetic beads spiked in a cell mixture and/or buffer sample.

2. Theories

2.1. Hydrodynamic focusing

In the symmetrical hydrodynamic focusing as shown in **Figure 1** below where $\mathbf{Q_{s1}}$ and $\mathbf{Q_{s2}}$ are equal, the sample stream width can be focused, constrained and centered in the middle of the microchannel, using the two equal sheaths flow rates from the peripheral channel inlets.

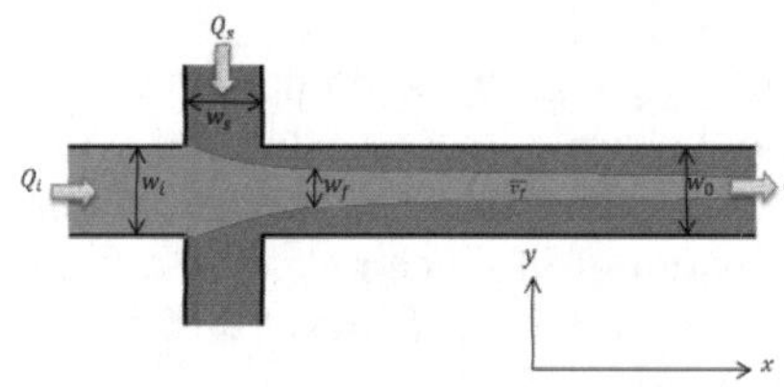

Figure 1. Hydrodynamic sample stream focusing mechanism

In the same manner this focusing can be determined regardless of the sample stream width location within the channel, by using an asymmetric hydrodynamic focusing, where the sheath flow rates $\mathbf{Q_s}$ and $\mathbf{Q_{s2}}$ are unequally injected. Here we only consume symmetrical focusing conditions. The width of hydro-dynamically focused stream ($\boldsymbol{w_f}$), relates to the inlet flow rates ($\boldsymbol{Q_i}$) and the two side channels flow rates (Q_S) as follow [18]

$$\frac{\mathbf{w_f}}{\mathbf{w_0}} = \frac{\bar{\mathbf{v}}_\mathbf{0}\, \mathbf{Q_i}}{\bar{\mathbf{v}}_\mathbf{f}\, (\mathbf{Q_i}+2(\mathbf{Q_s})} \tag{1}$$

Where
w_0 Outlet channel width, $\bar{v}_f$ and $\bar{v}_0$ are focused stream and outlet channel average flow velocities, respectively.
In order to determine the focused stream width, both $\bar{v}_f$ and $\bar{v}_0$ velocities, in equation (1) should be calculate as stated below
As viscous forces are predominant and the Reynolds number value is low in microfluidics, therefore solving the equation for a fully devolved flow in the outlet channel is straight forward. When applying wall no-slip boundary conditions consideration ,the solution can be expressed as [19].

$$\mathbf{u(y,z)} = \frac{4h^2}{\mu\pi^3}\left(-\frac{dp}{dx}\right)\sum_{n=0}^{\infty}(-1)^n \times \left\{1 - \frac{\cosh[(2n+1)\pi y/h]}{\cosh[(2n+1)\pi w_0/2h]}\right\}\frac{\cos[(2n+1)\pi y/h]}{(2n+1)^3} \quad (2)$$

Where
μ The fluid dynamic viscosity, $-w_0/2 \leq y \leq w_0/2$ and $-h/2 \leq z \leq h/2$ are the width and height of the outlet channel respectively.
Equation (2) is a rectangular channel with a Poiseuille velocity profile. To find $\bar{u}(y)$ the average velocity of the fluid, we can integrate equation (2) along z- direction using the previous boundary conditions

$$\bar{u}(y) = \frac{8h^2}{\mu\pi^4}\left(-\frac{dp}{dx}\right)\sum_{n=0}^{\infty}\frac{1}{(2n+1)^4} \times \left\{1 - \frac{\cosh\left[\frac{(2n+1)\pi y}{h}\right]}{\cosh\left[\frac{(2n+1)\pi w_0}{2h}\right]}\right\} \quad (3)$$

Using equation (3) both $\bar{v}_f$ and $\bar{v}_o$ velocities for the outlet and focused stream can be determined as

$$\bar{v}_f = 1 - \frac{192h}{\pi^5 w_f}\sum_{n=0}^{\infty}\frac{1}{(2n+1)^5}\frac{\sinh\left[\frac{(2n+1)\pi w_f}{2h}\right]}{\cosh\left[\frac{(2n+1)\pi w_0}{2h}\right]} \quad (4)$$

$$\bar{v}_0 = 1 - \frac{192h}{\pi^5 w_0}\sum_{n=0}^{\infty}\frac{\tanh\left[\frac{(2n+1)\pi w_0}{2h}\right]}{(2n+1)^5} \quad (5)$$

Equation (1) has been analytically solved for different (Q_s/Q_i) ratio values, particularly, 0 to 20, for a microfluidic channel with aspect ratio ($\varepsilon = h/w_0$) equal to 0.5.

2.2 Magnetic separation
A simple model was studied to separate magnetic particles and magnetically labelled cells from sample mixture.
A computational model using COMSOL Multiphysics software was devolved to find the magnetic field produced by the rectangular magnet located at pre-specified distance from the channel wall.

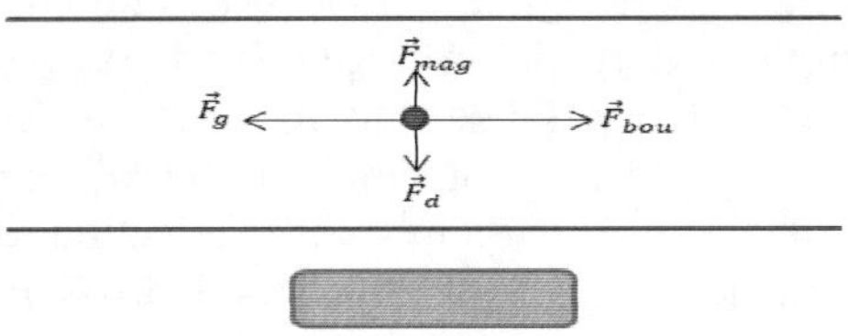

Figure 2. The four main forces act on a magnetic bead suspended in a liquid medium

Figure 2 above shows the main forces that acts the magnetic particle flowed in a liquid media, under effect of magnetic field, where $\vec{F}_d, \vec{F}_{mag}, \vec{F}_g$ and $\vec{F}_{bou}$are the drag, magnetic, gravity and the bouncy forces respectively. Here we just consider the magnetic and fluidic forces which are the most dominant forces.
The generated magnetic force on the magnetic particle can be given by[20]:

$$\vec{F}_{mag} = -\frac{4\pi r_p{}^3 \Delta x}{3\mu_0} (B.\Delta)B \quad (6)$$

Where

$\vec{B}$ Is the magnetic field, μ_0 is the magnetic permeability of free space, r_p magnetic particle radius and Δx is the bead, fluid magnetic susceptibility difference.

Here in our configuration, the y directions of the magnetic force, which leads to the deflection of magnetic particle and so the pre-labelled cells during their path was calculated as follow

$$\vec{F}_{mag,y} = \frac{4\pi r_p{}^3 \Delta x}{3\mu_0} B_y \frac{\partial B_y(x,y)}{\partial y} \quad (7)$$

2.3 The calculation of MNPs trajectory

The trajectories of the magnetic particles within the microfluidic device under magnetic force effects were achieved using force balance principle.

As mentioned previously, it was considered that the gravity and bouncy forces on the magnetic particles are quite small and can be neglected, comparing to the magnetic and viscous drag forces.

Using Stokes' approximation for the drag on a sphere, the hydrodynamic drag force ($\vec{F}_d$) can be expressed as

$$\vec{F}_d = 6\pi\eta r_p(\vec{v} - \vec{v}_p) \quad (8)$$

Where

η,$\vec{v}$ and $\vec{v}_p$ are the fluid viscosity, fluid and particle velocities respectively.The governed equation of the magnetic particle, given by

$$\vec{v}_{p,y} = \frac{dy}{dt} = \frac{\vec{F}_{mag,y}}{6\pi\eta r_p} \quad (9)$$

Equation (9) can be numerically solved; hence the magnetic particles trajectories can be determined.

3. Methodology

In our device, the hydrodynamic focusing has the ability to focus, align the particle and to orient cells into a narrow stream line located in the microchannel centre .Using this technique the particle manipulation can be achieve at a high throughput rate. As mentioned previously, the pre-labelled CTCs were firstly focused to a single narrow streamline using symmetrical hydrodynamic focusing technique, then in the second stage, the external magnetic applied field would alter and deflect the magnetic beads labelled CTCs from the focused particle streamline, thus the targeted cells could be separated with high efficiency from mixture of non-targeted cells sample or/buffer.

The above equations were utilized for magnetic bead focusing and tracing in 2D microfluidic channels, as shown schematically in figure 3 (A) below. The designed microchannel,s width (w) and length (L), used here in this analysis were 200μm and 30 mm in y and x direction respectively.

Firstly, in our study, the different flow rates, that affect the hydrodynamic focusing of the particles were investigated. Different flow rates ratios value of (Q_s/Q_i) were conducted, and the corresponding focused stream width was calculated analytically and numerically simulated used the above equations, furthermore, the focused stream widths, were experimentally measured.

In this study, the employed microchannel has three inlets and three outlets. The hydrodynamic focusing experiments were worked out using deionized water and deionized water-immersed 5 μm polystyrene beads as sheath and sample fluid respectively, followed into the microchannel using two different syringe pumps (LongerPump[R], TS-1B, China), in order to control the sheaths and sample flow rates precisely. Optical microscopy (OLYMPUS-IX7, Japan), was used to acquire the experimental hydrodynamic focusing effect images and analysis.

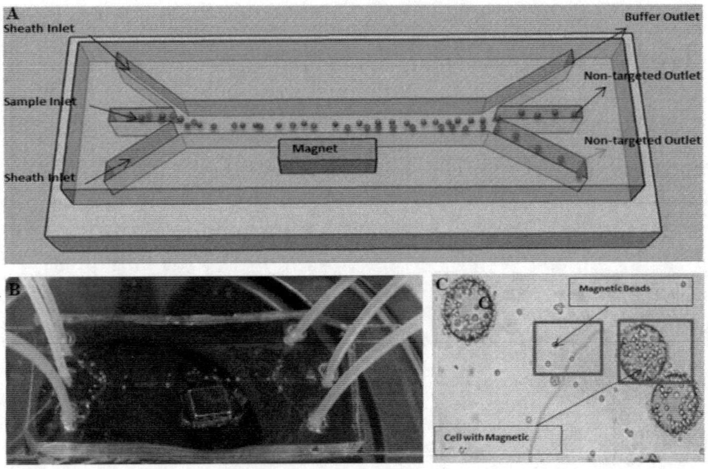

Figure 3. (A) Schematic shows the magnetic beads tracing. (B) The fabricated micro-device. (C) Anti-EpCAM coated beads and breast cancer cell line (MCF7) cells

Additionally, the microchannel was simulated numerically, using COMSOLMulti-Physics software; two dimensional simulations were carried out. Laminar flow, transport diluted species (tds), and particle tracing module (fpt) were the physics involved in this simulation. The sample channel flow rate was adjusted at10μl/min, while different sheath flow rates values were chosen according to the selected sample/sheath flow rate ratio values. The Full width at half maximum (FWHM) of the concentration cross-sectional plots was calculated and considered as the focused stream width $\boldsymbol{w_f}$ values, for the different conditions and adjusted sample/sheath flow rate ratios.
Hydrodynamic focusing helps in cell or particle alignment and focusing in a narrow single sample stream, after assuring of the hydrodynamic process works well, the magnetic separation process takes place.
To determine the magnetic field distribution in the configuration above, a 2 D model was constricted and simulated using Finite Element Method (FEM) package COMSOL. The simulation model includes the polydimethylsiloxane (PDMS) channel, the permanent magnet and the flow media, as shown in figure 3. In order to limit the simulation domain, outer circle was used. A 5 mm^3cubic shape and 49600 A.m saturation magnetization NdFeB magnet was used. This magnet was placed at 2 mm away from the channel wall.
Magnetic field strengthH_y in both channel walls (**at** $\boldsymbol{y = 0}$ and$\boldsymbol{H}$) was computed, so as the magnetic field gradient in y direction $\frac{\partial \boldsymbol{H_y(x,y)}}{\partial \boldsymbol{y}}$ throughout the microchannel width has been estimated to be 592.813 A/m.
We fabricated this device in order to separate the CTCs cells from buffer/sample based on magnetic method after aligning the cell sample into single narrow stream using hydrodynamic effect, super-magnetic beads coated with Anti-EpCAM antibodies were used to label MCF7 cells (taken here as a CTCs model), the MCF-7 cell line was purchased from American Type Culture Collection (ATCC). MCF7 were cultured in Eagle's Minimum Essential Medium (EMEM) with 0.01 mg ml−1 bovine insulin (Sigma-Aldrich) and 10% fetal bovine serum (FBS, Gemini Bio Products) as supplements. Trypsin-EDTA (Invitrogen) was used for MCF-7 cells harvesting, and then cells were re-suspended in media. Haemocytometer with microscope were used for cells counting, the average of three calculations used here as cells number.
To conjugate the magnetic beads into the desired antibody, a mixture of 10 μL from Anti-EpCAM antibody (Abcam) added to 20 μL from streptavidin modified magnetic bead (1 μm in diameter) for one hour at $27C^0$, then the complex was washed three times with PBS and column magnet, lastly anti-EpCAM beads were suspended in PBS buffer. The mixture of Anti-EpCAM coated beads and breast cancer cell line (MCF7 cells) showed a very high binding of the beads to the cells surface as shown in Figure 3(C).

4. Results and Discussion

4.1 Hydrodynamic focusing

As the aspect ratio in microchannel, is low, the velocity cross the micro channel width has a parabolic profile, using analytical solution of equation (3), gives the average flow velocity, which has been studied for different aspect ratio **ε** values. The results for the fully developed flow in the microchannel, given bellow in figure 4. As we can see from the graph, the average velocity field profile shape is totally based on the aspect ratio of the channel .Here the velocity field profiles for 0.1, 0.2, 0.5, 1 and 1.5 aspect ratios were calculated analytically using equation (3). One can clearly observe the transient behaviours of the velocity profile, between parabolic and plug-like profile during increasing and decreasing the aspect ratio respectively.

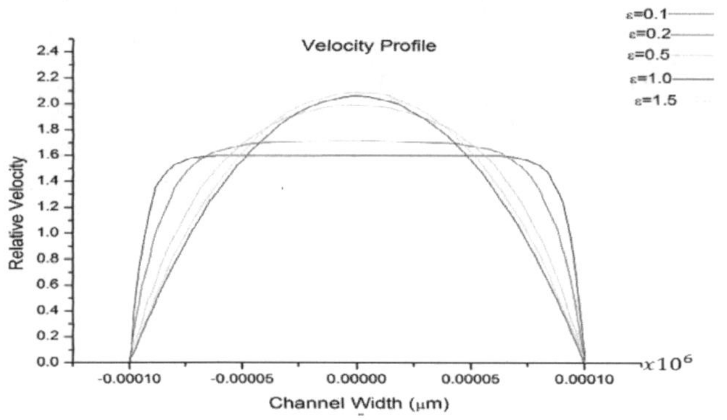

Figure 4. Velocity profiles for 0.1, 0.2, 0.5, 1.0 and 1.5 aspect ratio (ε) microchannel.

Firstly, as we mentioned previously, we studied the hydrodynamic focusing of the core sample stream under different conditions. The 0 to 20 values ($\boldsymbol{Q_s}$ =10-200 μl/min) of the flow rate ratios (Q_s/Q_i)has been studied, for the 0.5 aspect ratio microchannel. Particularly, by solving the pre-stated equations analytically, additionally, we compared the results given by analytical solution with that from the numerical and experimental one.

figure 5(A) below, shows examples of the calculated focused stream widths values for image sequences at (2,4,6,12,16,and 20) (Q_s/Q_i) flow rate ratio values respectively, and the method used in $\boldsymbol{w_f}$ calculations as FWHM of the cross-sectional concentration plots for COMSOL MULTIPHYSIC simulations figure 5(B). Additionally, the ratio of the collected particles during 100 seconds from both peripheral outlets after hydrodynamic focusing and the desired outlet after magnetic separation to the total number of the particle in the inlet (transmission probability), was calculated as shown in figure 7(A,B), in the simulation and practical process we used 30 and 10^4 particles in buffer solution respectively, using Dyna beads magnetic properties as stated by H Shirzadfar (2015)[21], the determination of the focused stream width w_f , was determined experimentally by utilizing deionized and deionized water-black ink mixture solution (1:5) as the two side and sample inlets respectively. Microscope integrated with a CCD camera ***(OLYMPUS- IX71)***, was used for flow images acquiring in figure 5(C), the images were analysed by ImageJ software, in order to get the light intensity of the cross-sectional plots. Again the FWHM of the light intensity cross-sectional plot cross the microchannel width was calculated.

Finally, using equation (1), the focused stream width $w_f$is computed analytically using MATLAB for the same different flow rate ratios values, the results has been tabulated in table 1 below.

The results of the focused stream width obtained, using the three methods has been summarized in figure 6, we can conclude that, all the three methods are perfectly matched with each other in focused width estimations and calculation. Furthermore, it clearly showed that the focused stream width is correlate strongly (linearly decrease) up to certain flow rate ratio value. As the flow rate ratio increases, the changes in the focused stream width no longer occurred or a little change happened, since non-linear correlation has been observed in case of high flow rate ratios values.

IOP Conf. Series: Materials Science and Engineering **350** (2018) 012013 doi:10.1088/1757-899X/350/1/012013

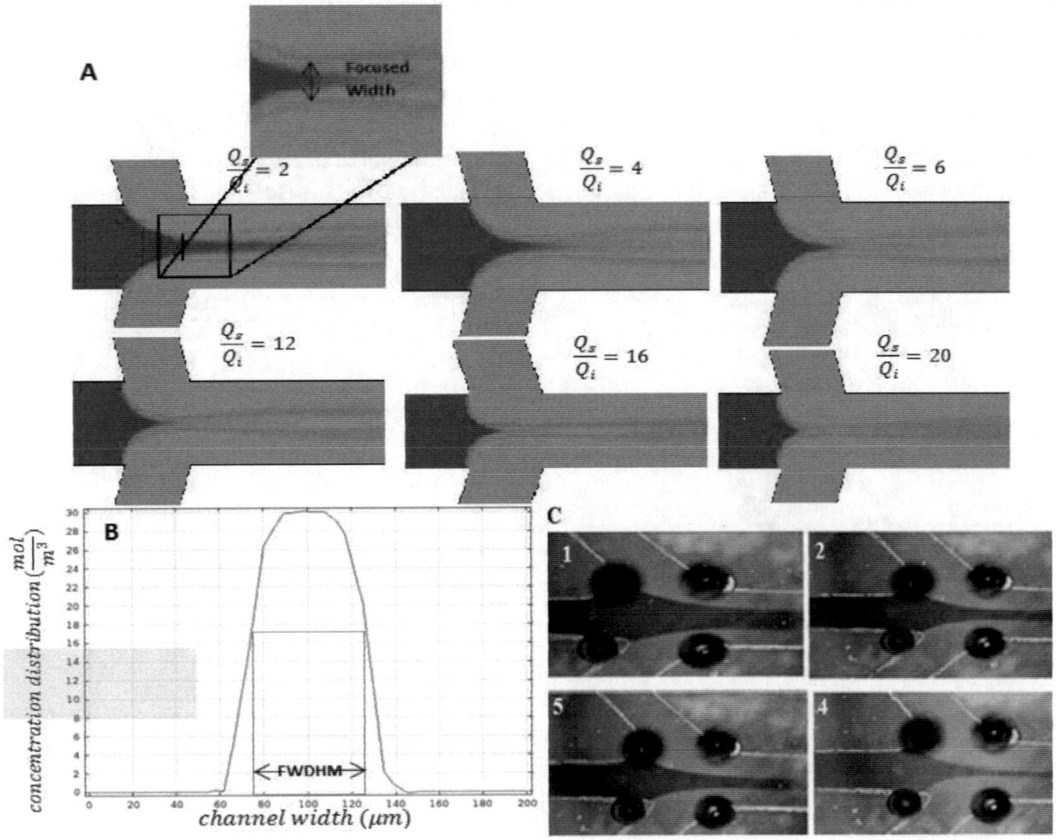

Figure 5. (A) Focused stream width images sequences from Comsol Multiphysics simulation. (B) Plotted curve example of the sample stream concentration along the width and the FWHM calculation method. (C) The selected experimental images (1, 2, 3 and 4) were obtained when $\frac{Q_s}{Q_i}$ values are 2, 4, 6 and 20 respectively.

Table 1. Analytically calculated w_f results.

$\frac{Q_s}{Q_i}$	1	2	4	6	8	10	12	14	16	18	20
$w_f(\mu m)$	50.494	29.8553	16.5316	11.4349	8.7411	7.0743	5.9418	5.1218	4.5006	4.0139	3.6222

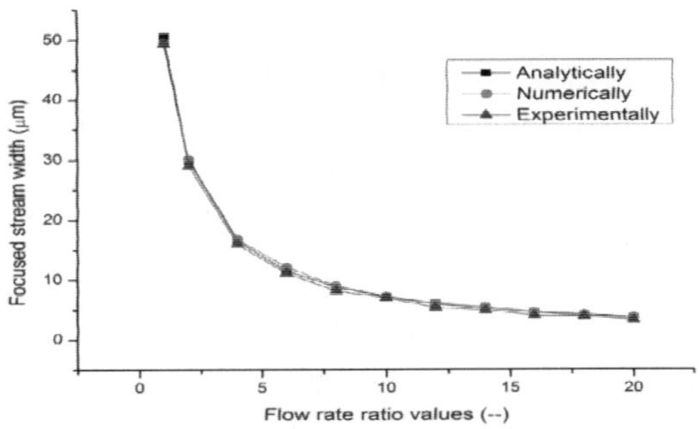

Figure 6. The results of the focused stream width obtained, using the three methods.

Particle trajectories for both focusing and magnetic simulations for time saving, was simulated only in the main channel domain. This domain has the same dimension parameters of the channel (200 um and 30 mm) high and length respectively. Figure 7(A) below showed the main device working characterization concept. For hydrodynamic focusing our desire is to statistically evaluate the particles that have been collected in outlet B (middle) with respect to the total particle from the inlet (B), while after magnetic separation the same analysis will be done for the particle collection from outlet (A)
The numerical particle focusing which has been conducted using comsol multiphysics results, has been analysed and used to calculate the *Transmission Probability.*

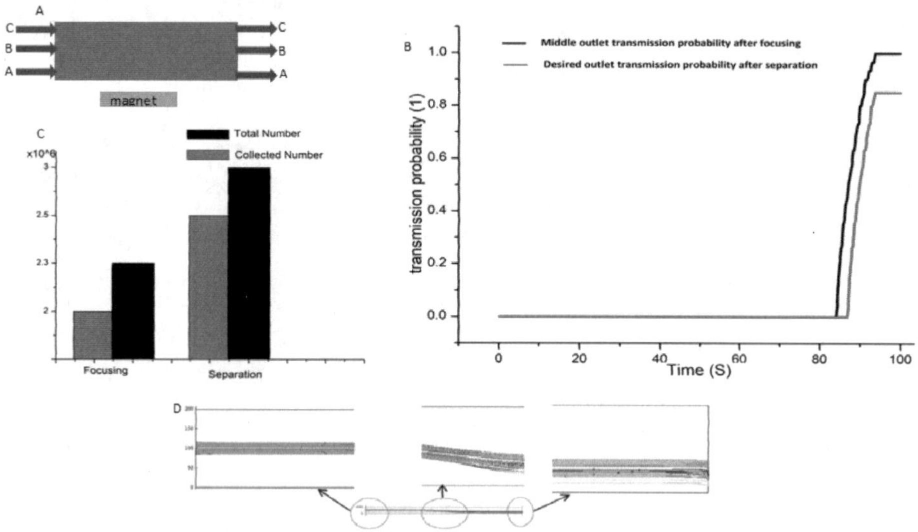

Figure 7. (A) schematically graph used for characterizations. (B) Transmission Probability after hydrodynamic focusing and magnetic separation a t$\frac{Q_s}{Q_i}$=4. (C) Transmission Probability experimental results. (D) Magnetic beads trajectories simulation at the inlet, middle (magnet zone) and outlet.

Which gives the ratio of the collected to the total particle number per time, as in figure 7 (B) (black line) we can figure out how efficient hydrodynamic focusing is, since within 82 to 85 second the particle reach the exit (B) (middle one). Additionally, *Transmission Probability* value reached 1 indicating 100% of particles that entering the channel exit out through the desired outlet (B)(middle), hence, 100% focusing efficiency achieved. We also verified these numerical results with an experimental one which resulted in 96 % focusing efficiency, by using known particle concentrations samples and counting the collected from the middle outlet using hemocytometer device. These differences can be returned to imprecision on the collected particles counting and concentration determination or particle stacking in the microchannel walls.

4.2 Magnetic separation
After assuring the hydrodynamic process, as used here to facilitate the cells separation by cell focusing and aligning in a narrow stream works well, the magnetic separation of the pre-magnetic beads labelled cells will take place, firstly a 2 dimensional simulation on COMSOL MULTIPHYSICS, has been constricted to study the magnetic field produced by the magnetic configuration used, beside studying magnetic force that acts on the magnetic beads and the trajectories have been investigated. X and Y components of the magnetic field gradient distribution, have been plotted at three different distances (1000, 1500 2100 um) from the magnet. In addition, we have calculated the y component of the magnetic force that acting on the magnetic particle trajectories based on equation (7) along the horizontal line at the middle of the microchannel (2100um) from the magnet), the results shown in figure 8. It's clearly that the Y – component of the Magnetic field gradient has negative values,

IOP Conf. Series: Materials Science and Engineering **350** (2018) 012013 doi:10.1088/1757-899X/350/1/012013

indicating the attractive force that deflects the magnetic beads during their path toward the desire outlet.

Additionally, the correlation between the magnetic field gradient and the corresponding generated force acts on the bead illustrated in figure 8 (B, C), shown direct relation as stated in equation (7).

Transmission Probability, as calculated for hydrodynamic focusing, also calculated her after magnetic separation as in figure 7(B) red line for outlet (A), as well as experimentally verified in figure 7(C), the simulations and experimental results shown a little bit different 92% and 95% respectively, which can be justified to the same reasons mentioned above, in addition to magnetic particles trapping near to the magnet zone in the microchannel.

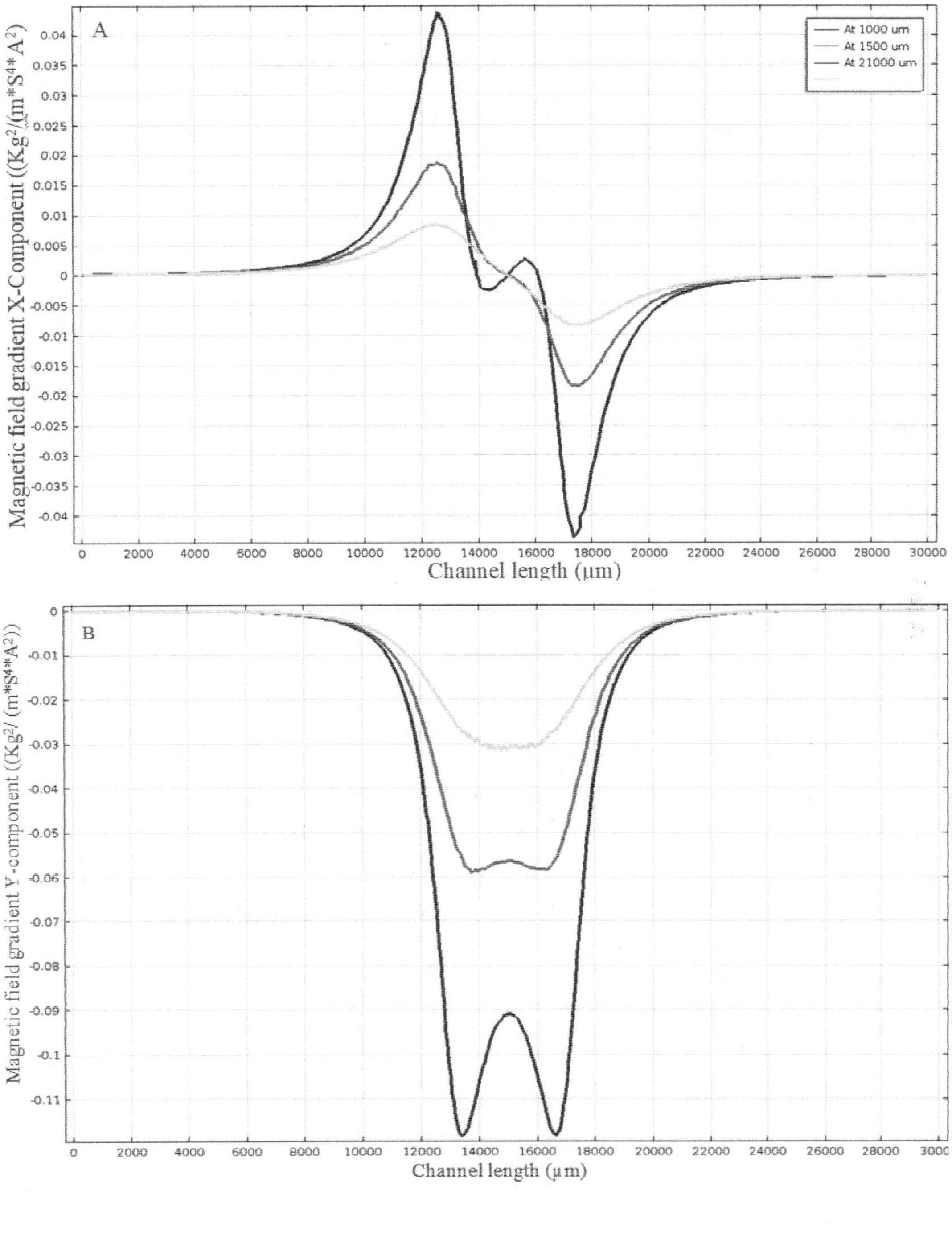

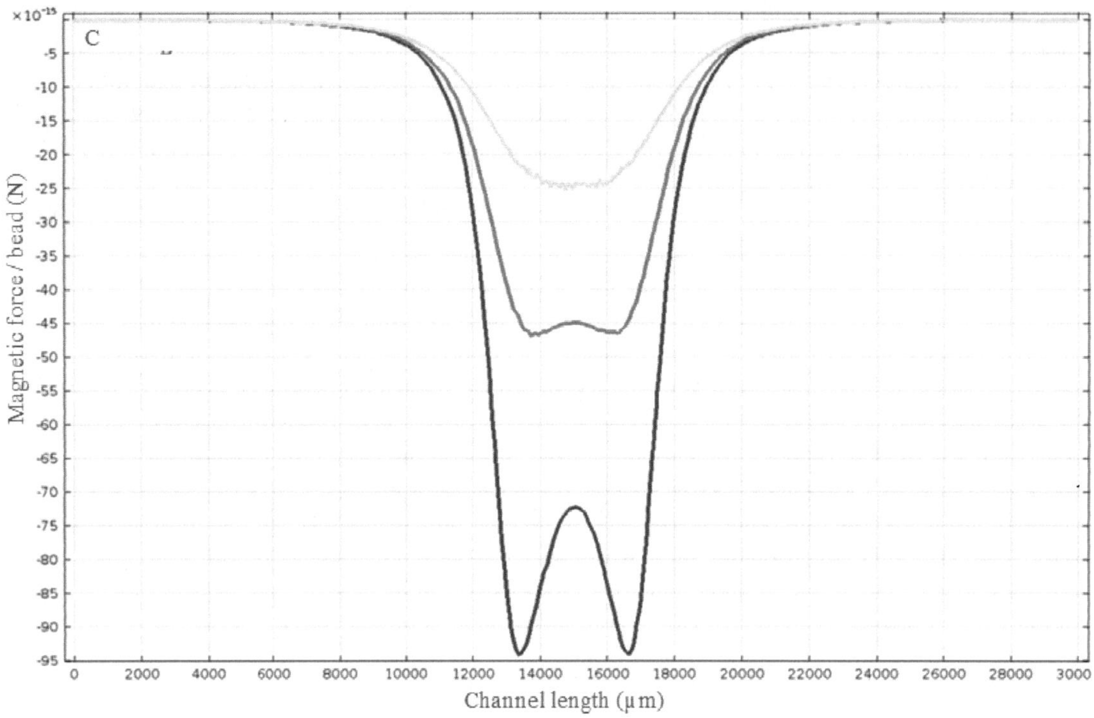

Figure 8. (A) magnetic field gradient X-component. (B). magnetic field gradient Y-component. (C). magnetic force per bead.

To simulate magnetic pre-labelled cells movement in COMSOL MULTIPHYSICS, firstly, we have to calculate the magnetic bead- cell complex magnetic susceptibility, which is totally different from that of the magnetic bead alone. Here we considered the combinations of the bead to the cell has no effect on the overall drag force, as the diameter of the bead considerably has negligible effect on the complex volume. Using the given bead data sheet information[22], volume, mass and ion content mass was calculated as 11.489 μ^3, 16.083x10−15kg and 1.930x10−15kg respectively, the same parameters have been used for cell as calculated by[23].The magnetic bead susceptibility and calculated cell mass with and without protein have been used to calculate the complex susceptibility as (0.087) using the following expression:

$$\mathbf{x_c = x_p * icont} \tag{10}$$

Where x_c , x_p and $icont$ are the complex, bead susceptibility and ion content respectively.

For practical magnetic cell separation tests, we prepared 500 μL white blood cells (WBCs) (2*10^3 cells), in phosphate buffer saline (PBS) after 0.1 % bovine serum albumin (BSA) adding, spiked with 100 MCF-7 cells. Cells were injected into the microchannel through sample inlet at fixed 10 $\mu L/min$ flow rate value and a solution of PBS contained 0.1% BSA was used as the two sheath fluids through the two sheaths inlets.

The microfluidic device characteristics were quantitatively estimated, using fluorescent microscope images as shown in figure 9, in which EpCAM, CD45, and DAPI are used to characterize MCF-7 from WBCs. Figure 9(A,B) shows the fluorescent images of the sample mixture before the separation and the fluorescent images of the collected cells sample.

A parameterized study for different spiked MCF7 cell number was carried out; the corresponding isolated number from the collected sample has been calculated. Also the purity was observed under fluorescent microscope, by counting the total number of the isolated MCF-7 cell and the contaminated

WBCs founded in the collected sample, as 98 and 5 cells respectively, as in figure 8(A,B,C) below, using the flowing expression:

$$\textbf{Purity} = \frac{C_{iso}}{C_{tot}} \tag{11}$$

Where:

C_{iso} And C_{tot} are the isolated and total numbers of MCF-7 respectively.

The MCF7 Cell separation purity using our device was 92.65 %. Additionally, the recovery rate for different spiked number of MCF7 (10-500) cells in 500 μl PBS a 0.1 % BSA buffer solution was calculated as shown in figure 8 (D).

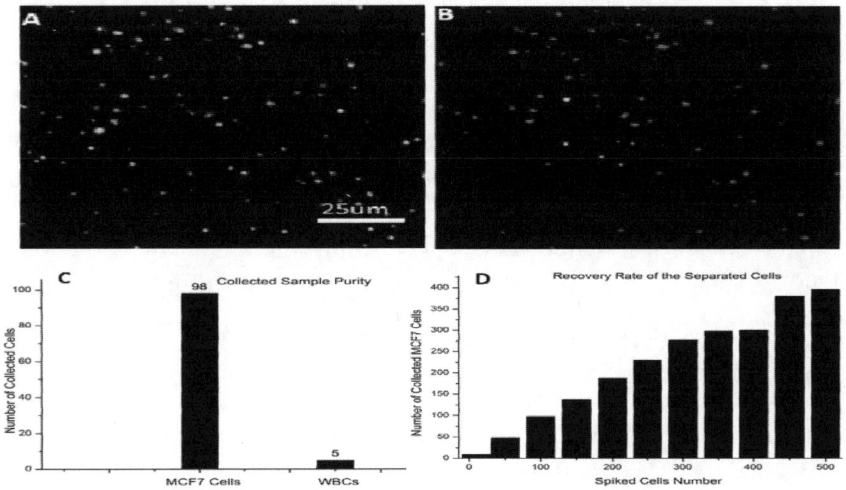

Figure 9. Microfluidic device Characterisations (A, B): fluorescent images of the sample mixture before the separation and the fluorescent images of the collected cells sample. (C): purity. (D): Recovery rate of the separated cells.

5. Conclusions

In this paper, we described the theoretical principles behind both hydrodynamic and magnetic phenomena utilized in microfluidic channel, along with experimental verifications of these theories. From the results given here, it obvious that compared to our previous paper which used only magnetic based cell[24], using hydrodynamic as an assistance tool has good impacts on the overall separation process enhancing. As our results for hydrodynamic focusing shows the efficiency of the hydrodynamic technique in particle as well as in cell focusing, as we get almost 100% transmission probability , indicating that almost all the particles have been focused and exited the microchannel at the desired middle outlet. And the overall particle loss was almost zero, which is clearly shown in figure 7. Additionally, we have clarified the high possibility given by hydrodynamic techniques, in focused stream width controlling, as it's a vital issue in cell separating and sorting for flow-cytometer devices fabrication. The effects of the sheath flow rates (Qs1 and Qs2), on the sample stream focusing have been studied, that to get the optimal conditions for the proposed device effectiveness and its capability in cell focusing. In the microchannel geometry used here, we found the optimal flow rate ratio, which gives the best fitted cells size (MCF7 Here, Dia 15.6 um) as being 6, at this flow rate ratios, the corresponding focused stream width was 16.5316 um, considerably in the same order of the cell size. At this end, the cells are aligned and sorted in a single narrow beam. Here the most obvious benefit is avoiding the cell accumulation and aggregation in specific region throughout the microchannel. The numerical simulations of the magnetic field gradient and the related force, gave fully consistent results with the analytical expression. *Transmission Probability concept,* has been calculated after each steps in our devise, additionally this concept used for the characterization of the fabricated device. Our combined hydrodynamic- magnetic microdevice, has shown it's highly ability in magnetic cell separation (MCF7 cells). As experimentally, we achieved more than a 96% recovery rate of MCF7 cells separation from a white blood cells mixture.

Our fabricated device is based on easy and cheap manufacturing process and operations procedure, which are the main requirements for clinical application instruments. Combining two or more techniques into microfluidic platforms, are promising solution for the limitations microfluidic chip cell separation.

6. Acknowledgements

This work in the school of life science, Beijing institute of technology and financially was funded by the National Key Scientific Instrument and Equipment Development Project of China (No. 2012YQ040140).

7. References

[1] E. W. K. Young and D. J. Beebe, "Fundamentals of microfluidic cell culture in controlled microenvironments.," *Chem. Soc. Rev.*, vol. 39, no. 3, pp. 1036–1048, Mar. 2010.

[2] P. R. C. Gascoyne and J. Vykoukal, "Particle separation by dielectrophoresis," *Electrophoresis*, vol. 23, no. 13. NIH Public Access, pp. 1973–1983, Jul-2002.

[3] C. W. Shields Iv, C. D. Reyes, and G. P. López, "From chip-in-a-lab to lab-on-a-chip: towards a single handheld electronic system for multiple application-specific lab-on-a-chip (ASLOC)," vol. 15, 2014.

[4] S. J. Tan, L. Yobas, G. Y. H. Lee, C. N. Ong, and C. T. Lim, "Microdevice for the isolation and enumeration of cancer cells from blood," *Biomed. Microdevices*, vol. 11, no. 4, pp. 883–892, Aug. 2009.

[5] C. Wyatt Shields IV, C. D. Reyes, G. P. López, D. Guo, C. Shen, W. P. Wong, K. Halvorsen, O. C. Farokhzad, G. S. Teo, J. A. Phillips, D. M. Dorfman, R. Karnik, J. M. Karp, L. Malaquin, J. L. Viovy, H. A. Wakelee, S. X. Wang, L. V. Sequist, R. J. Lee, K. J. Isselbacher, S. Maheswaran, D. A. Haber, M. Toner, A. Bardia, L. V. Sequist, D. N. Louis, S. Maheswaran, R. Kapur, D. A. Haber, and M. Toner, "Microfluidic cell sorting: a review of the advances in the separation of cells from debulking to rare cell isolation," *Lab Chip*, vol. 15, no. 5, pp. 1230–1249, Feb. 2015.

[6] Siyang Zheng, Raylene Yung, Yu-Chong Tai, and H. Kasdan, "Determinstic lateral displacement mems device for continuous blood cell separation," in *18th IEEE International Conference on Micro Electro Mechanical Systems, 2005. MEMS 2005.*, pp. 851–854.

[7] S. L. Feng, A. M. Skelley, A. G. Anwer, G. Liu, and D. W. Inglis, "Maximizing Particle Concentration in Deterministic Lateral Displacement Arrays."

[8] A. Armakolas, Z. Panteleakou, A. Nezos, A. Tsouma, M. Skondra, P. Lembessis, N. Pissimissis, and M. Koutsilieris, "Detection of the circulating tumor cells in cancer patients," *Futur. Oncol.*, vol. 6, no. 12, pp. 1849–1856, Dec. 2010.

[9] V. Zieglschmid, C. Hollmann, and O. Böcher, "DETECTION OF DISSEMINATED TUMOR CELLS IN PERIPHERAL BLOOD," *Crit. Rev. Clin. Lab. Sci.*, vol. 42, no. 2, pp. 155–196, Jan. 2005.

[10] R. N. Zare and S. Kim, "Microfluidic Platforms for Single-Cell Analysis," *Annu. Rev. Biomed. Eng.*, vol. 12, no. 1, pp. 187–201, Jul. 2010.

[11] H. Zhu, X. Lin, Y. Su, H. Dong, and J. Wu, "Screen-printed microfluidic dielectrophoresis chip for cell separation," *Biosens. Bioelectron.*, vol. 63, pp. 371–378, 2015.

[12] F. Petersson, L. Åberg, A.-M. Swärd-Nilsson, and T. Laurell, "Free Flow Acoustophoresis: Microfluidic-Based Mode of Particle and Cell Separation," *Anal. Chem.*, vol. 79, no. 14, pp. 5117–5123, Jul. 2007.

[13] M. Kersaudy-Kerhoas, R. Dhariwal, M. P. Y. Desmulliez, and L. Jouvet, "Hydrodynamic blood plasma separation in microfluidic channels," *Microfluid. Nanofluidics*, vol. 8, no. 1, pp. 105–114, Jan. 2010.

[14] Y. Gao, Y. Zhu, Z. Zhang, C. Zhang, X. Huang, and Z. Yuan, "Clinical significance of pancreatic circulating tumor cells using combined negative enrichment and immunostaining-fluorescence in situ hybridization," *J. Exp. Clin. Cancer Res.*, vol. 35, no. 1, p. 66, Dec. 2016.

[15] W. Sheng, O. O. Ogunwobi, T. Chen, J. Zhang, T. J. George, C. Liu, and Z. H. Fan, "Capture, release and culture of circulating tumor cells from pancreatic cancer patients using an enhanced mixing chip.," *Lab Chip*, vol. 14, no. 1, pp. 89–98, Jan. 2014.

[16] A. Truini, A. Alama, M. G. Dal Bello, S. Coco, I. Vanni, E. Rijavec, C. Genova, G. Barletta, F. Biello, and F. Grossi, "Clinical Applications of Circulating Tumor Cells in Lung Cancer Patients by CellSearch System.," *Front. Oncol.*, vol. 4, p. 242, 2014.

[17] O. O. Saeed, R. Li, and Y. Deng, "Microfluidic Approaches for Cancer Cell Separation: Review," *J. Biomed. Sci. Eng.*, vol. 07, no. 12, pp. 1005–1018, Oct. 2014.

[18] C. Kunstmann-Olsen, B. D. James Hoyland, and B. Horst-Günter Rubahn, "Syddansk Universitet Influence of geometry on hydrodynamic focusing and long-range fluid behavior in PDMS microfluidic chips Influence of geometry on hydrodynamic focusing and long-range fluid behavior in PDMS microfluidic chips."

[19] G.-B. Lee, C.-C. Chang, S.-B. Huang, and R.-J. Yang, "The hydrodynamic focusing effect inside rectangular microchannels," *J. Micromechanics Microengineering*, vol. 16, no. 5, pp. 1024–1032, May 2006.

[20] P. Chen, Y.-Y. Huang, K. Hoshino, and J. X. J. Zhang, "Microscale Magnetic Field Modulation for Enhanced Capture and Distribution of Rare Circulating Tumor Cells," *Sci. Rep.*, vol. 5, no. 1, p. 8745, Aug. 2015.

[21] H. Shirzadfar, M. Nadi, D. Kourtiche, S. Yamada, and P. Shahabi, "Characterization of a needle-type giant magnetoresistance sensor for detection of Escherichia coli's magnetic marker," *Int. J. Smart Sens. Intell. Syst.*, vol. 8, no. 1, pp. 220–234, 2015.

[22] "Magnetic Handling of Your Target 12."

[23] C. Fumarola, S. La Monica, R. R. Alfieri, E. Borra, and G. G. Guidotti, "Cell size reduction induced by inhibition of the mTOR/S6K-signaling pathway protects Jurkat cells from apoptosis," *Cell Death Differ.*, vol. 12, no. 10, pp. 1344–1357, Oct. 2005.

[24] O. Saeed, R. Li, and D. Yulin, "Micromagnetophoreses based CTCs Detections in Simple Chip."

IOP Conf. Series: Materials Science and Engineering **350** (2018) 012014 doi:10.1088/1757-899X/350/1/012014

Influence of Frequency-Dependent Dielectric Loss on Electrorheology of Surface Modified ZnO Nanofluids

H M Zaid[1], M Adil[1], KC Lee[1] and N R A Latiff[1]

1 Fundamental and Applied Sciences Department, Universiti Teknologi PETRONAS, 32610 Seri Iskandar, Perak Darul Ridzuan, Malaysia
E-mail: hasnamz@utp.edu.my

Abstract. The shear dependent viscosity change in dielectric nanofluids under the applied electric field, provide potentials for prospect applications especially in enhanced oil recovery. When nanofluids are activated by an applied electric field, it behaves as a non-Newtonian fluid under electrorheological effect (ER) by creating the chains of nanoparticles. In this research, the effect of dielectric loss on the electrorheological characteristic of dielectric nanofluids (NFs) was studied, corresponding to the applied frequency of 167 and 18.8 MHz. For this purpose, electrorheological characteristics of ZnO (55.7 and 117.1 nm) nanofluids with various nanoparticles (NPs) concentration (0.1, 0.05, 0.01 wt. %) were measured. The measurement was done via solenoid based EM transmitter under salt water as a propagation medium. The result shows that the applied electric field caused an apparent increase on the relative viscosity of ZnO NFs due to electrorheological effect. However, the relative viscosity shows a higher increment at 167 MHz due to the greater dielectric loss, compared to 18.8 MHz. The high dielectric loss allows the dipole moments to rotationally polarize at the interfaces of nanoparticles, which create stronger chains that align with the applied electric field. Additionally, the relative viscosity demonstrated an increment with the increase in particle size of ZnO nanoparticles from 55.7 to 117.1 nm. While the viscosity of nanofluid also indicated the high dependence on particle loading.

1. Introduction

Electrorheology is a field of science that investigates the viscosity of liquids, which changes reversibly and continuously under the impact of an electric field. The manipulation of the viscosity of electrorheological fluids (ERFs) under the effect of an applied voltage is called electrorheological (ER) effect [1-3]. The electrorheological fluid is a suspension of nano-size particles (such as ZnO, Al_2O_3) in a base fluid, which shows the enhancement in viscosity by the interconnection of particles to each other to form an ordered chain under the influence of an electric field. The direction of these chains is consistent with the direction of the force field; hence, increases the flow resistance of the liquid phase. This ER effect is also called the Winslow effect, after the name of the American inventor Willis Winslow who was the first researcher to investigate and published this phenomenon in 1949 [4].

Because of their unique characteristics, the ER fluid behaves as a functional fluid and has been successfully applied in damping devices, electrically controlled valves, micro robotics, and in aerospace engineering [5, 6]. Apart from their use in hydraulic systems, the ERFs provide a promising application in the upstream oil industry, especially for enhanced oil recovery (EOR) [7]. Therefore, the quantitative understanding of ER effect becomes important in the application of the nanofluid in the field of petroleum engineering. Parthasarathy and Klingenberg [8] have developed a polarization model to describe macroscopic behaviour of ER fluids. The polarization model suggested that the dielectric properties of dispersions, specifically dielectric loss is believed to control theelectrorheological effects. This model was further verified by Hao [9], which concluded that the nanofluids with a larger dielectric loss exhibited a stronger ER effect.

Considering the significant effect of electrorheology and suspended nanoparticles on the viscosity of nanofluids, it seems that the investigation of ER effect on viscosity and flow characteristics of nanofluids is an interesting subject while research in this area is limited. Sheng and Wen [3] explored the interaction between nanoparticles and an electric field from the electrorheological point of view. Raykar et al. [10] also presented the electrorheological properties of low-concentration Fe_2O_3 nanoparticles suspended in ethylene glycol under the influence of electric fields. Also, it was seen that the use of stabilizer (acetylacetone) helps in nanoparticles dispersion properties and is responsible for the electrorheological effect. While, Yin and Zhao [11] reported the recent researches on electrorheology of several nanofiber-based suspensions, including organic, inorganic, and inorganic/organic composite nanofibers. These electrorheological fluids can also be manipulated as an EOR agent by improving the viscosity of displacing fluid, which is one of the EOR mechanism. Haroun et al. [7] also investigated the enhance oil recovery by employing the copper- and nickel oxide nanoparticles (50 nm) in the presence of an external electric field. They termed it as electrical EOR (EEOR), which shows an improved oil recovery by 9 – 22 % over the recovery from waterflooding.

However, a critical issue which could not be ignored in the course of the examination of electrorheology is the problem of ensuring the coverage of the applied electric field due to the large well spacing in an oil field. Hence, despite the presence of the stimulating electric field, the corresponding frequency is crucial to achieve an electrorheological phenomenon. In the previous study [12], the viscosity of dielectric nanofluids under electromagnetic (EM) field have been investigated under different propagation media at a lab-scaled frequency of 167 MHz (corresponding to 1000 m of field-scaled distance). The results showed that the intensity of the ER effect does not only depend on the shear rate, and electric field strength; but also strongly dependent on microstructure and agglomeration of nanoparticles in basefluid with a surfactant. The present work is an effort to investigate the influence of electrorheological effect on viscosity of ZnO nanofluids at a lab-scaled frequency of 18.8 MHz (corresponding to 3000m of well spacing), in comparison to 167 MHz. The effect of particle concentration of nanofluid (ZnO/SDBS) on electrorheology is also analyzed under the electric field in the presence of salt water (≈ seawater concentration) as an EM propagation medium.

2. Experimental

2.1. Materials

ZnO nanoparticles were synthesized using sol-gel auto-combustion method [13], and were used after calcining at 500 and 800°C having an average particle size of 55.7 and 117.1 nm respectively. In this paper, these particles were denoted as ZnO@500 and ZnO@800, respectively. The analytical grade of sodium dodecylbenzenesulfonate (SDBS) from Sigma Aldrich was used as the stabilizer, without further purification. Deionized water (with σ = 18 MΩ) was used as a solvent. NaCl obtained from Fisher Scientific, was employed as salt to prepare brine of a concentration of 30000 ppm (equivalent to sea water concentration). The pH value of system was adjusted with HCl and NaOH solution by precise pH meter (Mettler Toledo, FE20-Basic).

2.2. Nanofluid Preparation

The nanoparticles were dispersed in brine as the base fluid and magnetically stirred for 1 hour to produce nanoparticles suspension. Then, the appropriate amount of sodium dodecylbenzenesulfonate (SDBS) was added to the suspensions. These suspensions were agitated in an ultrasonic bath at ambient temperature for an optimum period, to attain the required concentration of nanofluids. The anionic surfactant, SDBS, was chosen as a stabilizer based on our previous stability tests; where the surfactant concentrations were selected using critical micelle concentration (CMC) determination methods. The pH value of the system was also adjusted by using HCl and NaOH solution to improve the quality of dispersion. These pH values were monitored by precise pH meter (FE20-Basic) from Mettler Toledo.

2.3. Dielectric Properties
The dielectric properties of ZnO nanoparticles, calcined at different temperatures, were determined by using RF impedance analyzer (Agilent 4291B). The impedance analyzer is connected with Agilent 16453A Dielectric Material Test Fixture, as shown in Fig. 1, at the frequency range of 1 MHz to 1 GHz was used, to measure the dielectric properties of calcined nanostructures. The nanoparticles' pellets of 18mm diameter and ~2 mm thickness were formed by applying a pressure of about 50 MPa using a hydraulic press. These pellets are then utilized for the measurement of dielectric properties; including real ε' and imaginary ε'' part of the permittivity, along with the tangent loss. The structure of the dielectric fixture, as shown in figure 1, is similar to the "parallel plate capacitor" and consists of an upper electrode (with a diameter of 10 mm) and a bottom electrode (with a diameter of 7 mm). The upper electrode has an internal spring, which allows the pellet to be fastened between the electrodes. Applied pressure can be adjusted as well. It should be noted that silver (paste) is highly recommended to use to minimize the errors due to air gaps.

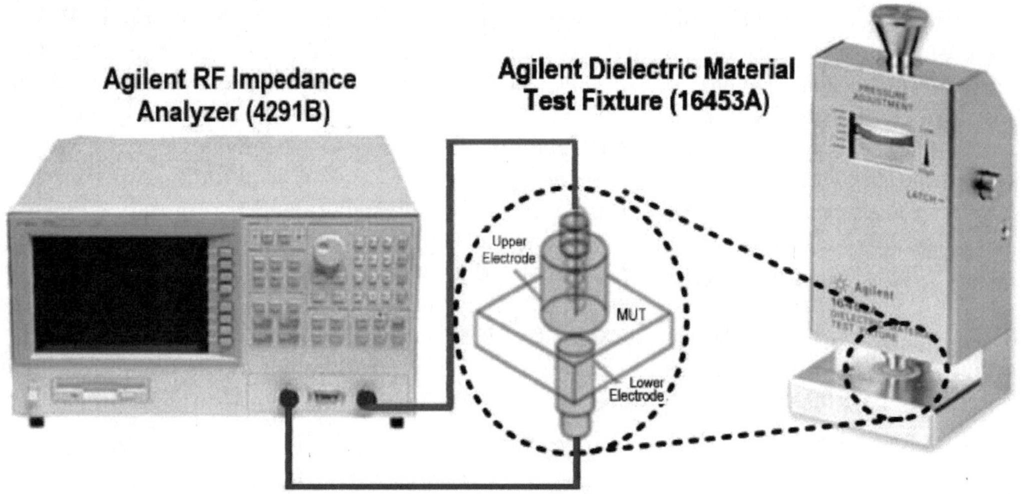

Figure 1. Setup of equipment for dielectric properties measurement

2.4. Electrorheology of ZnO Nanofluids
To investigate the effect of frequency on electrorheology of ZnO nanofluids, nano-ZnO samples with varying solid contents (0.1, 0.05, and 0.01 wt. %) were prepared, along with optimal stability parameters. A rotating viscometer (Brookfield DV-I+) attached to a custom-built solenoid coil was used to measure the viscosity of nanofluid. A separate solenoid coil was used for each designed frequency of 167 and 18.8 MHz, having a diameter of 14 and 42 cm respectively. The schematic diagram of measuring setup is depicted in figure 2.

The electric field perpendicular to the stainless steel (SS) sample chamber was generated by the solenoid surrounding the spindle. The UL adapter spindle was chosen to be used with the viscometer, which required 16 ml of nanofluid for every measurement. The suggested torque during the experimental measurement was maintained within 10–90% of the maximum torque. However, the low-viscous fluids cannot be measured in this range, especially at low shear rate. Therefore, the viscosity measurements at low shear rates were conducted in less than 10% torque with more measurement points and average technique to have better accuracy, as recommended by Behi and Mirmohammadi [14].

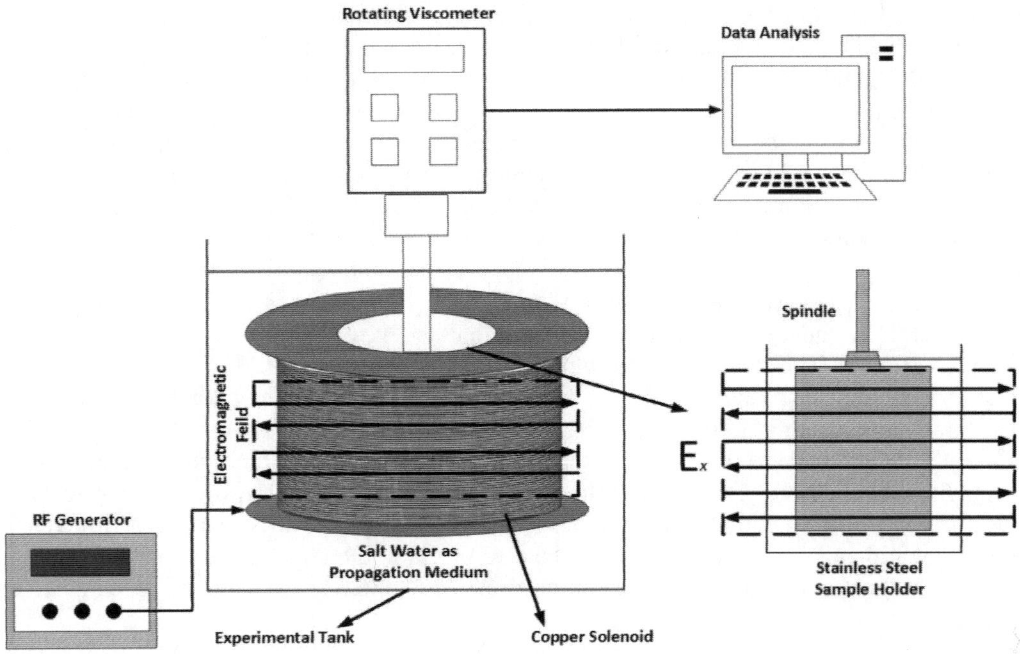

Figure 2. Schematic representation of experimental setup for measuring nanofluid viscosity under applied EM field.

3. Results and Discussion

3.1. Dielectric Measurements

The dielectric properties of ZnO are dependent upon various factors namely, chemical composition, method of preparation, grain size, etc. Fig. 3 shows the plots of ε', ε'' and tan δ of the ZnO nanoparticles as a function of calcining temperature. It is clearly shown that the typical trend at the applied selected frequencies is the increase of dielectric properties but with different rates according to the calcination temperature and the applied frequency. This can be described by the fact that polarization is the size dependent property (for bigger particles the polarization will be more). On the other hand, it has been narrated by some researchers [15, 16], that the size of the conducting grains increases with temperature i.e. the intergranular spacing decreases. Based on the particle size distribution, the particle size of ZnO keep on increasing with the increase in calcination temperature from 500°C to 800°C, thereby decrease the intergranular spacing of the non-conduction regions. Hence increase the dielectric polarization, which in turn increases the both parts of permittivity and the tangent loss.

The change in dielectric properties can also be verified by investigating the effect of applied frequencies, based on different polarization mechanisms which are contributing to the enhanced dielectric behavior of nanomaterials. We believe that rotational/orientational polarization process is the major contributor corresponding to the dielectric loss of the ZnO nanoparticles at the applied frequencies. For a conventional n-type semiconductor, the large amount of oxygen vacancies acts as shallow donors in ZnO. These oxygen vacancies exist in the interfaces of ZnO nanoparticles [17, 18]. Positive oxygen vacancies combine with the negative oxygen ions produce a significant amount of dipole moments. These dipole moments rotate in an external electric field, leading to the occurring of rotational polarization at the interfaces of n-type ZnO nanoparticles. However, at the low frequency of 18.8 MHz, the ZnO particles show a decrease in the dielectric loss which makes the dipolar groups hard to orient at the similar pace as the alternating field. Therefore, the influence of these dipolar groups to the permittivity goes on decreasing at the low frequency.

IOP Conf. Series: Materials Science and Engineering **350** (2018) 012014 doi:10.1088/1757-899X/350/1/012014

3.2. Frequency-Dependent Electrorheological Properties

The electrorheological characterization of ZnO nanofluids has been conducted under the presence of EM waves at designed frequency of 167 and 18.8 MHz, propagated in salt water. The relative viscosity of ZnO@500 nanofluid, as depicted in figure 4, shows a shear-thinning behavior for both the frequencies of 167 and 2V (figure 4a) and 18.8 MHz and 3.5V (figure 4b). It is found that the applied electric field had an apparent increase on the relative viscosity of ZnO NFs due to electrorheological effect. As indicated by figure 4, the applied voltage of 2V at 167 MHz had a greater effect on the relative viscosity of ZnO@500 NFs, than the relative viscosity at 18.8 MHz. At 167 MHz, the relative viscosity at 7.34 s^{-1} for ZnO@500 NPs (0.1 wt. %) recorded to be 3.25, compare to 2.4 at 18.8 MHz with 3.5V of applied voltage. This can be explained by the fact that the viscosity of NFs is determined by the dielectric properties of ZnO nanoparticles, where the viscosity of basefluid was not affected by an applied electric field. Under the electric field, dielectric nanoparticles are polarized and the dipole moments rotate corresponding to the dielectric loss of the nanoparticles. This leads to the occurring of rotational/orientational polarization at the interfaces of nanoparticles, which create chains that align with the applied electric field. However, at the low frequency of 18.8 MHz, the ZnO particles show a decrease in dielectric loss (figure 3) which makes the dipolar groups less responsive to even a high applied electric field at 3.5V. This makes them hard to orient at the equal pace as the alternating field. Therefore, the influence of these dipolar groups to the permittivity goes on decreasing at the low frequency.

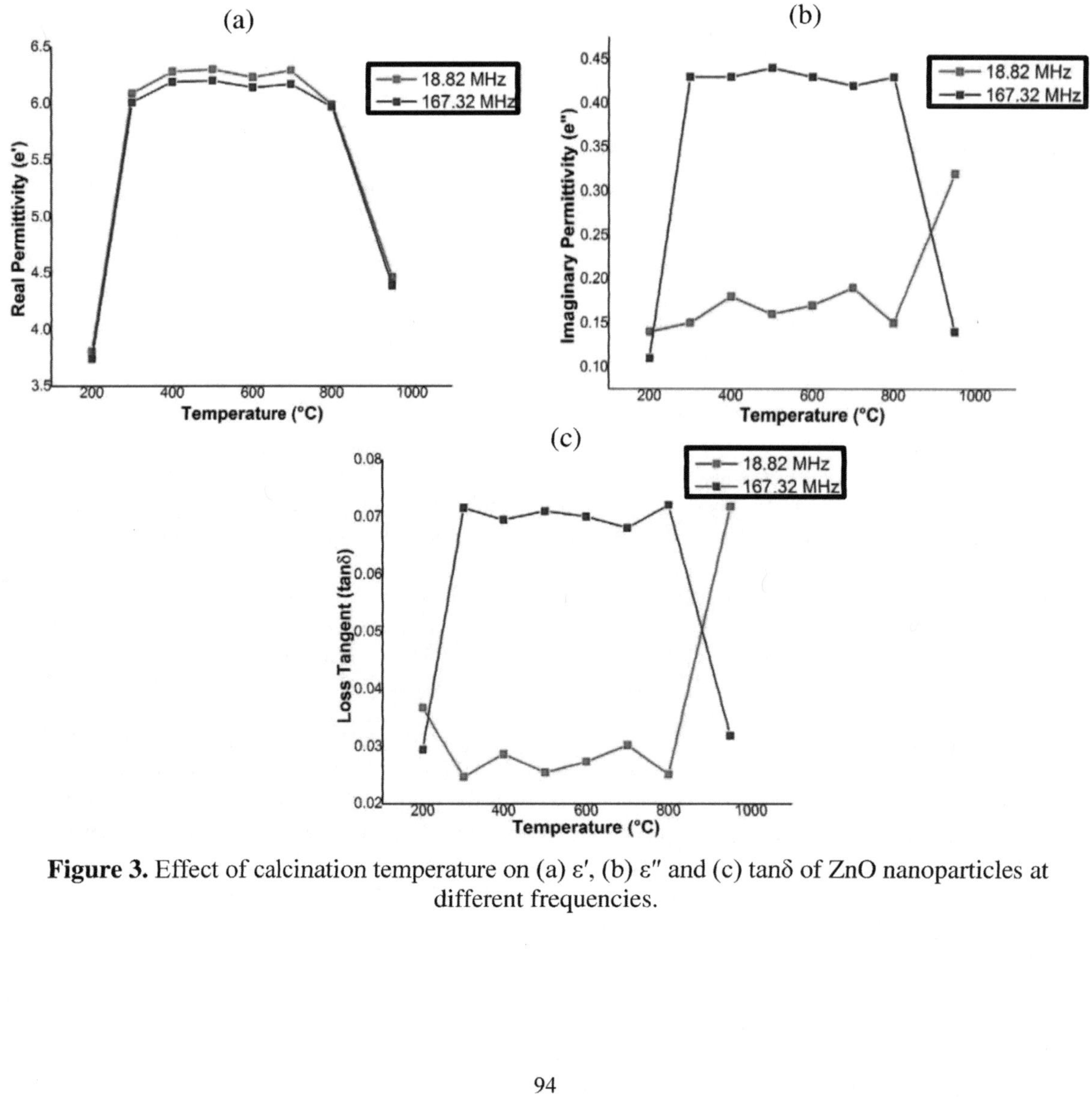

Figure 3. Effect of calcination temperature on (a) ε′, (b) ε″ and (c) tanδ of ZnO nanoparticles at different frequencies.

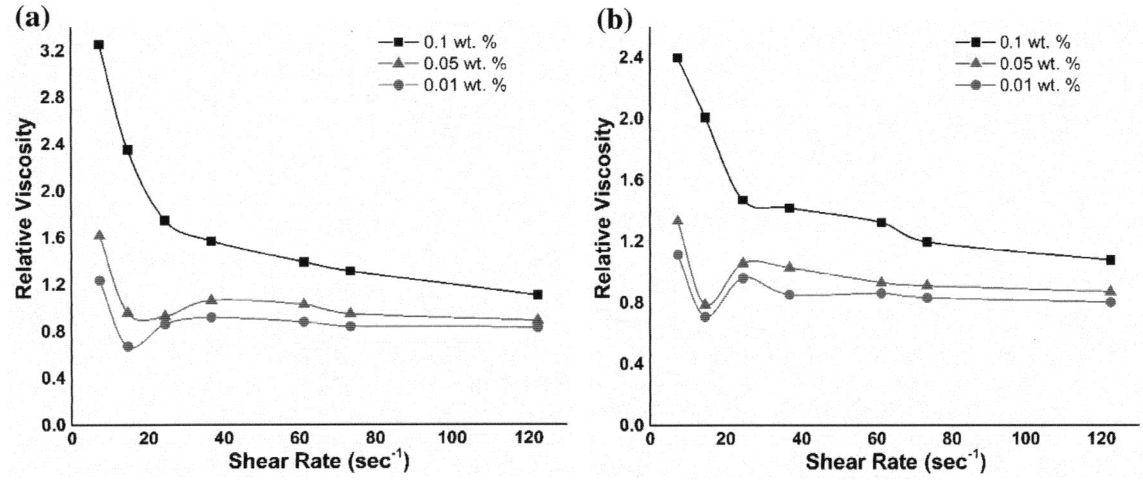

Figure 4. Relative viscosity versus shear rate of ZnO@500 nanofluids at different particle concentrations, under EM waves at (a) 167 MHz & 2V, and (b) 18.8 MHz & 3.5V.

Whereas in the case of ZnO@800 NFs, as shown in figure 5, the particle size plays a significant role; where the relative viscosity increases with the increase in particle size from 55.7 to 117.1 nm. This increment is in accordance with the investigation performed by Sinha et al. [19]. Here, the percentage increase in low-shear relative viscosity of 0.1 wt. % ZnO@800 NPs is recorded to be 3.9 and 3.33 at 167 and 18.8 MHz, respectively. However, at a maximum share rate of 122.4 s^{-1}, the dipolar force between the rotational particles diminished. Therefore, the chains cannot be maintained, the structure of NFs is destroyed, and the relative viscosity decreases noticeably to 1.14 and 1.11 at 167 and 18.8 MHz, respectively.

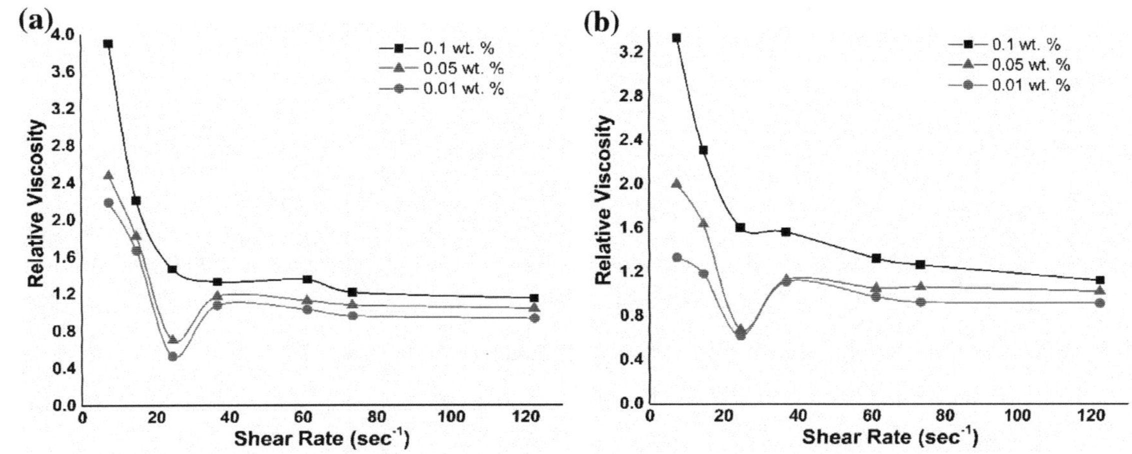

Figure 5. Relative viscosity of ZnO@800 nanofluids under electromagnetic waves at an applied frequency and voltage of (a) 167 MHz & 2V, and (b) 18.8 MHz & 3.5V.

3.3. Effect of Particle Concentration

By comparing the curves in figure 4 and figure 5, it was also observed that the nanofluids with higher concentration (0.1 wt.%) of nanoparticles exhibit obvious higher relative viscosity than of lower concentrations (0.05 and 0.01 wt.%). In the previous study [20], it was established that the possible existence of particle aggregations at the lower concentration of ZnO nanoparticles could be responsible for the decrease in relative viscosity. Thus, ZnO@500 & ZnO@800 NFs (0.1 wt. %) under electric field at 167 MHz presents a greater relative viscosity of 3.25 & 3.9 at the share rate of 7.34 s^{-1}; which is significantly higher than 1.61 & 1.23 and 2.47 & 2.19 for ZnO NFs at 0.05 and 0.01 wt. %, respectively. Similarly, in the case of electrorheology at 18.8 MHz (3.5 V), the decrement in relative viscosity was observed by the decrease in particle concentration.

4. Conclusion
The electrorheological characteristics of ZnO non-Newtonian nanofluids were investigated experimentally, corresponding to MHz frequency of 167 and 18.8. The relative viscosity at 7.34 s^{-1} for 0.1 wt. % ZnO@500 and ZnO@800 nanofluids show a greater increment of 3.25 and 3.9 respectively at 167 MHz and 2V. This is due to the higher dielectric loss at 167 MHz, compared to 18.8 MHz which lead to a comparatively smaller increment of 2.4 and 3.33 in relative viscosity even with a relatively high E-field strength at 3.5V. This suggests that the dipolar groups of ZnO at 18.8 MHz find it hard to orient at the similar pace as the alternating E-field. Therefore, the influence of signal frequency on these dipolar groups is critical to achieve a better electrorheological effect. However, at a maximum share rate of 122.4 s^{-1}, the dipolar force between the rotational particles diminished; which decreases the relative viscosity of ZnO@500 noticeably to 1.10 and 1.07 at 167 and 18.8 MHz, respectively. While, the relative viscosity of ZnO@800 at high share is measured to be 1.14 at 167 MHz, and 1.11 at 18.8 MHz. These results also clearly depicted that the particle size has a significant effect on the viscosity of nanofluids. On the other hand, the nanofluids with higher concentration (0.1 wt. %) of nanoparticles demonstrated a higher viscosity than those of lower concentrations (0.05 and 0.01 wt. %). In conclusion, the nanoparticles should have suitable dielectric properties to achieve a sustainable ER effect under steady shear rate.

5. Acknowledgments
The authors would like to acknowledge Ministry of Higher Education Malaysia (FRGS Grant No: 0153AB–I47) and Department of Fundamental and Applied Sciences, Universiti Teknologi PETRONAS for funding the research.

6. References
[1] Wen W, Huang X, Yang S, Lu K and Sheng P, 2003 The giant electrorheological effect in suspensions of nanoparticles, Nat. Mater. **2** 727-30

[2] Jiang W, Jiang C, Gong X, and Zhang Z 2009 Structure and electrorheological properties of nanoporous $BaTiO_3$ crystalline powders prepared by sol–gel method, *J.Sol-Gel Sci. Technol.* **52** 8-14

[3] Sheng P and Wen W 2012 Electrorheological fluids: mechanism, dynamics, and microfluidics applications, *Annu. Rev. Fluid Mec.* **44** 143-74.

[4] Winslow WM 1949 Induced fibration of suspensions, *J. Appl. Phys.* **20** 1137-40

[5] Choi HJ and John MS 2009 Electrorheology of polymers and nanocomposites, *Soft Matter* **5** 1562-67

[6] Choi Y, Sprecher AF and Conrad H 1990 Vibration characteristics of a composite beam containing an electrorheological fluid, *J. Intell. Mater. Syst. Struct.* **1** 91-104.

[7] Haroun M, Hassan SA, Ansari A, Kindy NA, Sayed NA, Ali B and Sarma H 2012 Smart nano-EOR process for Abu Dhabi carbonate reservoirs, *Abu Dhabi International Petroleum Conference and Exhibition, Abu Dhabi, UAE.*

[8] Parthasarathy M, Klingenberg DJ 1996 Electrorheology: mechanisms and models, *Mat. Sci. Eng.* **R 17** 57-103

[9] Hao T 2002 Electrorheological suspension, *Adv. Colloid Interfac.* **97** 1-35

[10] Raykar VS, Sahoo SK and Singh AK 2010 Giant electrorheological effect in Fe_2O_3 nanofluids under low dc electric fields, *J. Appl. Phys.* **108** 034306

[11] Yin J and Zhao X 2011 Electrorheology of nanofiber suspensions, *Nanoscale Res. Lett.* **6** 256

[12] Adil M, Zaid HM, Lee KC and Latiff NRA 2016 Effect of EM propagation medium on electrorheological characteristics of dielectric nanofluids, *Journal of Dispersion Science and Technology.* **38** 570-6

[13] Adil M, Zaid HM, Lee KC and Latiff NRA 2015 Microscopic evolution of dielectric nanoparticles at different calcination temperatures synthesized via sol-gel auto-combustion, *AIP Conf. Proc.* **1669** 020016.

[14] Behi M and Mirmohammadi SA 2012 Investigation on thermal conductivity, viscosity and stability of nanofluids, M.Sc. Thesis, Royal Institute of Technology, Department of Energy Technology, Stockholm, Sweden

[15] Vila R, Gonzalla M and Ibarra A 1998 Dielectric spectroscopy of alumina ceramics over a wide frequency range, *J. Nucl. Mater.* **253** 141-8

[16] Govinda S and Rao KV 1975 Dielectric properties of single crystals of Al_2O_3 and Al_2O_3 doped with chromium or vanadium, *Phys. Status Solidi (A)* **27** 639-44

[17] Boguslawski P, Briggs EL and Bernholc J 1995 Native defects in gallium nitride, *Phys. Rev. B* **51** 17255–58

[18] Perlin P et.al 1995 Towards the identification of the dominant donor in GaN, *Phys. Rev. Lett.* **75** 296–9

[19] Sinha TA, Kumar A, Bhargava N and Mallick S 2014 An experimental investigation into the thermal properties of nano fluid, *J. App. Mech. Eng.* **4** 1000148

[20] Adil M, Zaid HM, Lee KC and Latiff NRA 2016 Effect of dispersion stability on electrorheology of water-based ZnO nanofluids, *Energy Fuels.* **30** 6169-77

IOP Conf. Series: Materials Science and Engineering **350** (2018) 012015 doi:10.1088/1757-899X/350/1/012015

TiO_2/water Nanofluid Heat Transfer in Heat Exchanger Equipped with Double Twisted-Tape Inserts

S. Eiamsa-ard[1], R. Ketrain[2] and V. Chuwattanakul[3]

[1]Faculty of Engineering, Mahanakorn University of Technology, Bangkok, Thailand
[2]Royal Irrigation Department, Samsen Nakornchaisri, Bangkok, Thailand
[3]Faculty of Engineering, King Mongkut's Institute of Technology Ladkrabang, Bangkok, Thailand
Corresponding e-mail: smith@mut.ac.th

Abstract. Nowadays, heat transfer enhancement plays an important role in improving efficiency of heat transfer and thermal systems for numerous areas such as heat recovery processes, chemical reactors, air-conditioning/refrigeration system, food engineering, solar air/water heater, cooling of high power electronics etc. The present work presents the experimental results of the heat transfer enhancement of TiO_2/water nanofluid in a heat exchanger tube fitted with double twisted tapes. The study covered twist ratios of twisted tapes (*y/w*) of 1.5, 2.0, and 2.5) while the concentration of the nanofluid was kept constant at 0.05% by volume. Observations show that heat transfer, friction loss and thermal performance increase as twist ratio (*y/w*) decreases. The use of the nanofluid in the tube equipped with the double twisted-tapes with the smallest twist ratio (y/w = 1.5) results in the increases of heat transfer rates and friction factor up to 224.8% and 8.98 times, respectively as compared to those of water. In addition, the experimental results performed that double twisted tapes induced dual swirling-flows which played an important role in improving fluid mixing and heat transfer enhancement. It is also observed that the TiO2/water nanofluid was responsible for low pressure loss behaviors.

1. Introduction

Several techniques have been applied for improving the heat transfer and thermal performance in thermal systems [1-11]. Using nanofluids instead of base or common fluids is one of the techniques for improving the heat transfer in many thermal engineering processes including power generation, chemical production, heat exchanger, cooling processes, air-condition, heating, transportation, microelectronics and chemical processes. In general, the common fluids (water, ethylene glycol, mineral oil, engine oil and transformer oil) possess poorer heat transfer properties than most solids. To improve heat transfer, solid particles in nanoscale known as nanoparticles are mixed with base fluids, forming "nanofluids". Nanofluids are considered as not only a passive technique, but also a promising way, which open a new window to the future of advanced thermal fluid science. Nanoparticles suspended in conventional fluids were extensively utilized in heat exchangers for improving their thermal performance such as Al_2O_3/water [6-7, 9, 12], CuO/water [9, 12, 15-17], and TiO_2-water [18-19]. For their studied range, the heat transfer coefficient monotonically increased with increasing nanofluid concentration. Although, the combined heat transfer enhancement techniques using nanofluid together with typical twisted tape have been extensively studied and reported as literature review shown above, the explanation for the effect of twisted tape architecture together with nanofluid and also nanofluid concentration on the thermal performance is still limited and scarcely reported.

This is the driving force for this research work. The present work proposes the combined technique consisting of the use of TiO_2/water nanofluids and double twisted tapes as the modified tube inserts. The thermal performance was evaluated by simultaneous determining heat transfer and friction loss. The study covered twist ratios of twisted tapes (*y/w*) of 1.5, 2.0, and 2.5) and Reynolds numbers (*Re*) from 5000 to 20,000.

Figure 1. SEM image of TiO_2 nanoparticles.

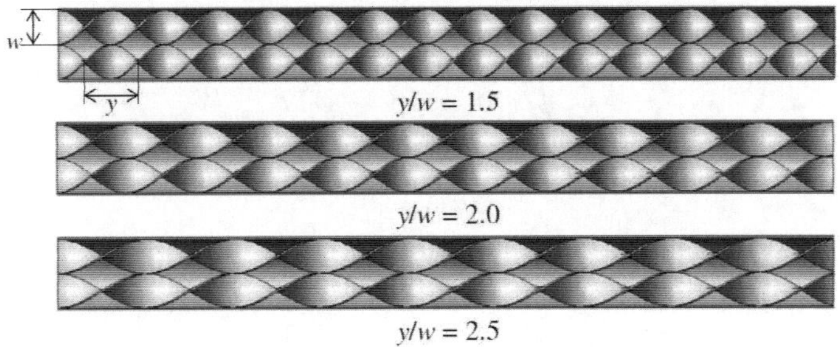

Figure 2. Double twisted tape with three twist ratios.

2. Experimental setup

2.1. TiO_2/water nanofluid

The nanoparticles used in the present work are AEROXIDE® TiO_2 P25, the product of Evonik Degussa. Figure 1 shows the scanning electron microscopy (SEM) photograph of the TiO_2 particles. Prior to testing, TiO_2 nanoparticles were dispersed in de-ionized water (the base fluid) at concentrations (ϕ) = 0.05% by volume. Then, the mixture was sonicated continuously for 180 minutes in an ultrasonic bath under ultrasonic pulses of 100 W at 36±3 kHz for the uniform dispersion of particles. Moreover, the sonication treatment significantly improved the stability of suspension. It took longer than 180 minutes after preparation for TiO_2 nanoparticles to start precipitation. Thus, TiO_2 nanoparticles were still well dispersed in water before feeding into the tube.

2.2. Double twisted tapes

Figure 2 demonstrated the geometries of the double twisted tapes. The tapes were made of thin aluminum sheets with a thickness of 0.8 mm and a width of 10 mm (*w*). Each twisted tape was fabricated by twisting a straight tape, about its longitudinal axis, while being held under tension. Twisted tapes were twisted at twist lengths (180°/twist length) of 15, 20 and 25 mm corresponding to twist ratios (*y/w*) of 1.5, 2.0, and 2.5, respectively. In the experiments, double twisted tapes were inserted into the inner tube of double-pipe heat exchanger.

Figure 3. Photograph of the experimental setup.

2.3. Data reduction

The thermos-physical properties of the nanofluid were calculated as a function of nanoparticle volume concentration (ϕ) together with properties of base fluid and nanoparticles. The density of nanofluid was evaluated using the simple formula for mixture:

$$\rho_{nf} = (1-\phi)\rho_{water} + \phi\rho_{np} \tag{1}$$

The specific heat of the nanofluid was evaluated from

$$c_{p,nf} = \frac{\phi\rho_{np}c_{p,np} + (1-\phi)\rho_{water}c_{p,water}}{\rho_{nf}} \tag{2}$$

The thermal conductivity was calculated as seen below

$$\frac{k_{nf}}{k_{water}} = \frac{k_{np} + 2k_{water} + 2\phi(k_{np} - k_{water})}{k_{np} + 2k_{water} - \phi(k_{np} - k_{water})} \tag{3}$$

Viscosity of nanofluids was calculated via the general Einstein's formula

$$\mu_{nf} = \mu_{water}(1+\eta\phi) \tag{4}$$

Nusselt number is calculated using the following equation

$$Nu = hD/k \tag{5}$$

The friction factor is written as following equation below

$$f = (D/L)(2\Delta P/\rho U^2) \tag{6}$$

The thermal performance factor (η) under same pumping power criteria is given by

$$\eta = (Nu/Nu_p)/(f/f_p)^{1/3} \tag{7}$$

3. Validation experiments of plain tube with water as the base fluid

Initially, the accuracies of the experimental data including heat transfer (Nu) and friction factor (f) were validated by comparing the present results (of distilled water as the base fluid in the plain tube) with the predictions of Dittus-Boelter and Gnielinski correlations for Nusselt numbers and those of Blasius and Petukhov correlations for friction factors. The comparisons are shown in Fig. 4. Apparently, the present data agree well with those of the standard correlations. This can be concluded that the results from the present experimental system are reliable. Thus, the facility was employed for further investigation with double twisted tapes and TiO_2/water nanofluid.

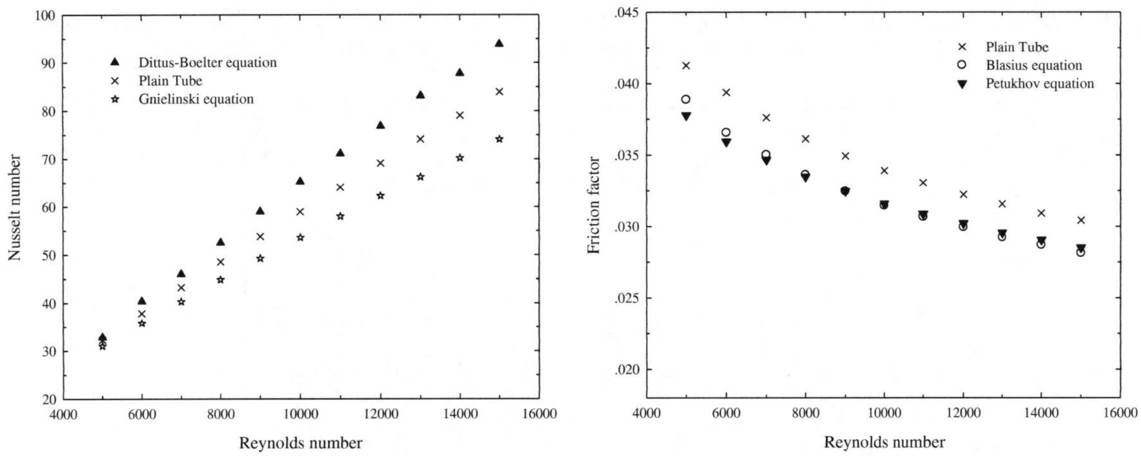

(a) Nu (b) f

Figure 4. Verification test of the present plain tube with water as base fluid.

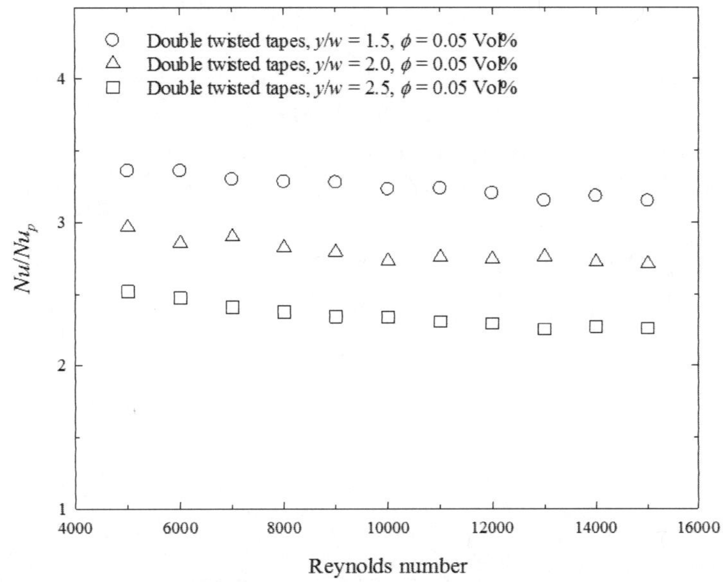

Figure 5. Influence of ϕ on Nu.

4. Results and discussion

4.1. Heat transfer

In this section, the effect of combined heat transfer enhancement technique using TiO_2/water nanofluids in double-pipe heat exchanger fitted with double twisted tapes on the heat transfer (Nu), friction factor (f) and thermal performance factor (η) are reported. Figure 5 depicts the variation of Nusselt number with Reynolds number for water and TiO_2/water nanofluid in the tube equipped with double twisted-tapes with twist ratios (y/w) = 1.5, 2.0 and 2.5. The Reynolds number was varied from 5000 to 20,000. Evidently, Nusselt number increases with increasing Reynolds number. At a given Reynolds number, all TiO_2/water nanofluids give higher Nusselt number than water as the based fluid. The higher heat transfer by nanofluids arises from: (i) the ability of suspended nanoparticles enhancing thermal conductivity; (ii) movement of nanoparticles delivering energy exchange. In addition, heat transfer increases as twist ratio of twisted tape decreases (or twist number increases). This can be attributed to the increase of swirling/turbulence intensity as well as the improved consistency of swirl flow which leads to thinner thermal boundary layer and better heat transfer. The use of the nanofluid in the tubes equipped with twisted tapes with *y/w* = 1.5, 2.0 and 2.5 resulted in heat transfer enhancement by 5.18%, 4.94% and 4.74%, respectively, compared to that of water.

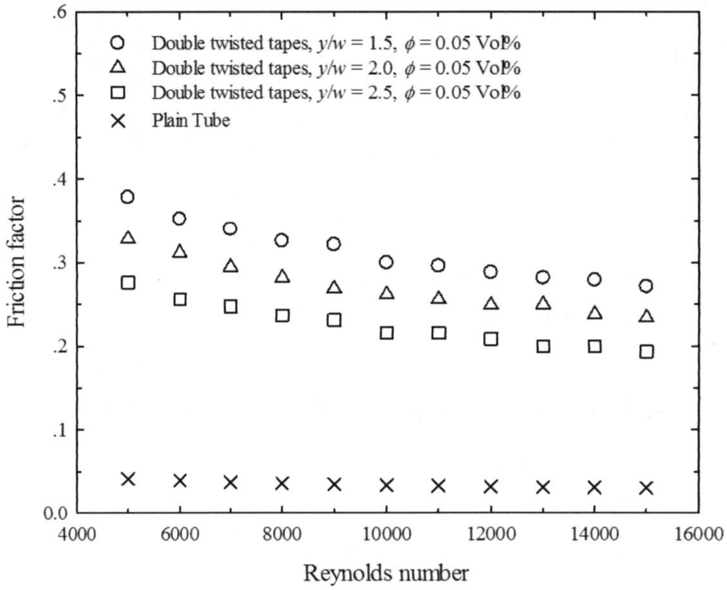

Figure 6. Influence of ϕ on *f*.

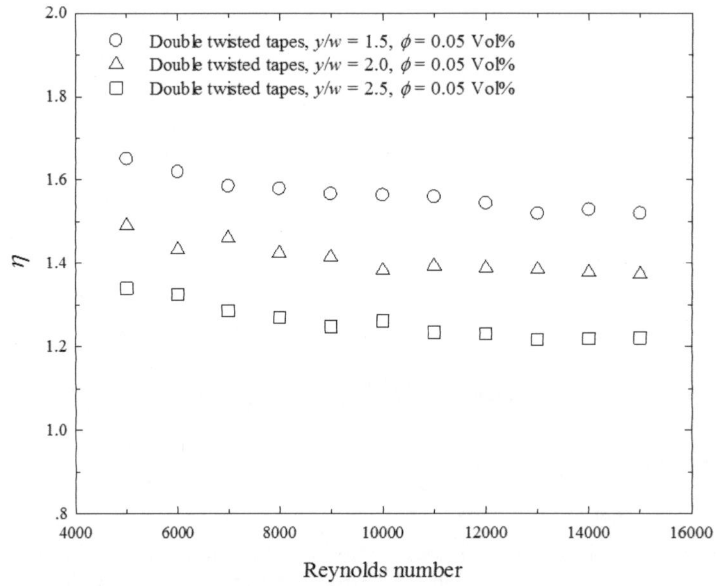

Figure 7. Influence of ϕ on η.

4.2. Friction factor

The effect of TiO_2/water nanofluids on the friction factor (*f*) in tube with double twisted-tapes is presented in Fig. 6. In general, the friction factor of tube with double twisted-tape inserts was higher than that the plain tube due to the swirling flow and the dissipation of dynamic pressure of the fluid at high viscosity loss near the tube wall. When twist ratio (*y/w*) decreases, friction factor increases. At a given Reynolds number, nanofluids cause higher friction factor than water (the based fluid). The rise in friction factor is due to the rise in viscosity of the nanofluids as compared to that of water. In addition, friction factor increases as twist ratio decreases. This result can be explained by the same reasons for heat transfer or Nusselt number.

4.3. Thermal performance factor

As shown above, the heat transfer enhancement technique increases both heat transfer (advantage) and friction loss (disadvantage). The overall performance or benefit can be evaluated by simultaneous determining heat transfer and friction loss in the form of thermal performance factor which is expressed in Eq. (8). Over the range investigated, the thermal performance factors of TiO_2/water nanofluid in the tubes with double twisted-tapes at y/w = 1.5, 2.0 and 2.5 are higher than that of the based fluid in the plain tube by around 56.6%, 41.1% and 25.9%, respectively, as seen in Fig. 7. The maximum thermal performance factors of TiO_2/water nanofluid in the tubes with double twisted-tapes at y/w = 1.5, 2.0 and 2.5 are 1.65, 1.49 and 1.33 are obtianed at Re = 5000.

5. Conclusion

The experimental results show that the use of nanofluid together with double twisted-tapes leads to the increases of heat transfer, friction loss and thermal performance as compared to those of the base fluid in the plain tube. Heat transfer, friction loss and thermal performance increase with decreasing twist ratio *(y/w)*. The Nusselt numbers of the nanofluid in the tubes equipped with the double twisted-tapes at y/w = 1.5, 2.0 and 2.5 increase by 224.8%, 179.6% and 134.8% over the base fluid in the plain tube. The highest thermal performance factor of 1.65 was obtained by the use of the nanofluid in the tube equipped with the double twisted-tapes at y/w = 1.5.

6. References

[1] Koolnapadol N, Sripattanapipat S and Skullong S 2014 *J. Res. Appl. Mech. Eng.* **4** 166
[2] Eiamsa-ard S, Yongsiri K, Nanan K and Thianpong C 2012 *Chem. Eng. Proc.: Proc. Intensification* **60** 42
[3] Matsunaga T and Sumitomo T 2014 *J. Res. Appl. Mech. Eng.* **2** 65
[4] Kongkaitpaiboon V, Nanan K and Eiamsa-ard S 2010 *Int. Comm. Heat Mass Transfer* **37** 560
[5] Nanan K and Eiamsa-ard P 2014 *J. Res. Appl. Mech. Eng.* **2** 103
[6] Sharma K V, Syam Sundar L and Sarma P K 2009 Int. *Comm. Heat Mass Transfer* **36** 503
[7] Syam Sundar L and Sharma K V 2010 *Int. J. Heat Mass Transfer* **53** 1409
[8] Suresh S, Venkitaraj K P, Selvakumar P and Chandrasekar M 2012 *Exp. Therm. Fluid Sci.* **39** 37
[9] Eiamsa-ard S, Somkleang P, Nuntadusit C and Thianpong C 2013 *App. Therm. Eng.* **54** 289
[10] Thianpong C, Eiamsa-ard P, Promvonge P and Eiamsa-ard S 2012 *Energy Procedia* **14** 1117
[11] Thianpong C, Yongsiri K, Nanan K and Eiamsa-ard S 2012 *Int. Comm. Heat Mass Transfer* **39** 861
[12] Suresh S, Venkitaraj K P and Selvakumar P 2011 *Superlattices and Microstructures* **49** 608
[13] Eiamsa-ard S 2010 *Int. Comm. Heat and Mass Transfer* **37** 644-651
[14] Promvonge P and Eiamsa-ard S 2007 *Int. Comm. Heat Mass Transfer* **34** 72
[15] Eiamsa-ard S and Wongcharee K 2012 *Int. Comm. Heat Mass Transfer* **39** 1453
[16] Wongcharee K and Eiamsa-ard S 2012 *Int. Comm. Heat Mass Transfer* **39** 251
[17] Wongcharee K and Eiamsa-ard S 2011 *Int. Comm. Heat Mass Transfer* **38** 742
[18] Eiamsa-ard S, Kiatkittipong K and Jedsadaratanachai W 2015 *Eng. Sci. and Tech., Int. J.* **18** 336
[19] Perarasu V T, Arivazhagan M and Sivashanmugam P 2012 *J. Hydro.* **24** 942

IOP Conf. Series: Materials Science and Engineering **350** (2018) 012016 doi:10.1088/1757-899X/350/1/012016

Effect of Capillary Tube's Shape on Capillary Rising Regime for Viscos Fluids

F Soroush[1] and A Moosavi[2]

1 Undergraduate Student, Sharif University of Technology, Tehran, Iran
2 Associate Professor, Sharif University of Technology, Tehran, Iran
E-mail: moosavi@sharif.edu

Abstract. When properties of the displacing fluid are considered, the rising profile of the penetrating fluid in a capillary tube deviates from its classical Lucas-Washburn profile. Also, shape of capillary tube can affect the rising profile in different aspects. In this article, effect of capillary tube's shape on the vertical capillary motion in presence of gravity is investigated by considering the properties of the displacing fluid. According to the fact that the differential equation of the capillary rising for a non-simple wall type is very difficult to solve analytically, a finite element simulation model is used for this study. After validation of the simulation model with an experiment that has been done with a simple capillary tube, shape of the capillary tube's wall is changed in order to understand its effects on the capillary rising and different motion regimes that may appear according to different geometries. The main focus of this article is on the sinusoidal wall shapes and comparing them with a simple wall.

1. Introduction

Although capillary rising is a phenomenon that is known from many years ago [1], but according to the fact that its new applications like manufacturing new detectors and measurement instruments are appearing more and more nowadays [2-4], it seems that the need for further researches about the effect of the different parameters on the capillary fluid motion in order to take the numerous advantages of this phenomenon and increasing our efficiency in using capillary-based instruments is increasing. Three main parameters which are less-known about vertical capillary rising, are effects of gravity, tube shape and viscosity of the displacing fluid. Some authors provided models for spontaneous imbibition [5], while some others investigated effects variably shaped apertures [6]. Recently, the effects of viscosity and gravity for simple wall capillary tubes are investigated [7]. In this study, after validation of our finite element model with an experiment, we are going to investigate the effects of changing tube's geometry on the vertical capillary rising, while considering both gravity and viscosity effects of fluids, which were neglected partially or completely in the past studies. It will be shown that the new geometry will affect the regime of motion in an especial way, and according to various applications of capillary tubes, this new phenomenon can be used in the future for different applications.

2. Governing equations

Equations that govern the physics of this study are the continuity equation (equation (1)) along with the Navier-Stoke's momentum equations for incompressible fluid (equation (2) to equation (4)) in all three dimensions.

Continuity equation:

$$\frac{\partial u}{\partial x} + \frac{\partial v}{\partial y} + \frac{\partial w}{\partial z} = 0 \qquad (1)$$

With u, v, w as the velocity magnitude in x, y and z direction, respectively.

Navier-Stokes equations:

$$x - component: \rho\left(\frac{\partial u}{\partial t} + u\frac{\partial u}{\partial x} + v\frac{\partial v}{\partial y} + w\frac{\partial w}{\partial z}\right) = -\frac{\partial p}{\partial x} + \mu\left(\frac{\partial^2 u}{\partial x^2} + \frac{\partial^2 u}{\partial y^2} + \frac{\partial^2 u}{\partial z^2}\right) + F_x \quad (2)$$

$$y - component: \rho\left(\frac{\partial v}{\partial t} + u\frac{\partial v}{\partial x} + v\frac{\partial v}{\partial y} + w\frac{\partial v}{\partial z}\right) = -\frac{\partial p}{\partial y} + \mu\left(\frac{\partial^2 v}{\partial x^2} + \frac{\partial^2 v}{\partial y^2} + \frac{\partial^2 v}{\partial z^2}\right) + F_y \quad (3)$$

$$z - component: \rho\left(\frac{\partial w}{\partial t} + u\frac{\partial w}{\partial x} + v\frac{\partial w}{\partial y} + w\frac{\partial w}{\partial z}\right) = -\frac{\partial p}{\partial z} + \mu\left(\frac{\partial^2 w}{\partial x^2} + \frac{\partial^2 w}{\partial y^2} + \frac{\partial^2 w}{\partial z^2}\right) + F_z + \rho g_z \quad (4)$$

Where , μ and are average density and viscosity , respectively. Also, F_x , F_y and F_z are magnitude of the surface tension force in Cartesian coordinate system.

3. Simulation

3.1. Finite element modelling

In order to model the physics, we used a finite element modelling program. According to the fact that the tube has a circular cross section, we used the axisymmetric solving, so the symmetry that exists in this problem can be used and calculation time reduces. An image of a simple capillary tube's mesh is shown in Figure 1. For the fluid flow, wetted wall boundary condition is applied for walls. In the outlet, the atmospheric pressure is set, and for the inlet, the pressure calculated and set according to hydrostatic fluid pressure law equation (5), where the density of the displacing fluid and its height is used in order to calculate the necessary pressure.

$$P = \rho g h \quad (5)$$

It is noteworthy that the motion of the penetrating liquid will not affect this pressure in the inlet, so there is no need to change it during the simulation.

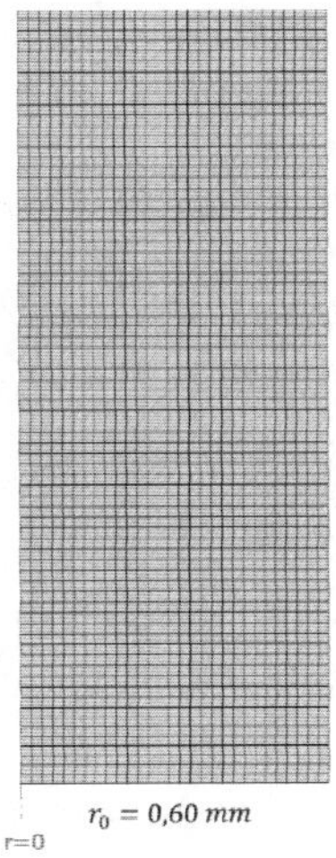

Figure 1. Simple tube mesh.

According to the experiment study [7], Oil (density 965 kg/m^3 and viscousity 97 mPa.s) and Glycerin (density 1225 kg/m^3 and viscousity 97 mPa.s) are set as the displacing and penetrating fluid, respectively.

3.2. Validation

Mesh independency has been investigated through checking the height of the capillary fluid in several different times for different mesh sizes and then, they compared with the results obtained from the experiment [7]. The optimum mesh that has been used in this study for different geometries, has an error less than 5% in comparison with the experiment results for each point. Results of experimental and simulation study are showed in Figure 2.

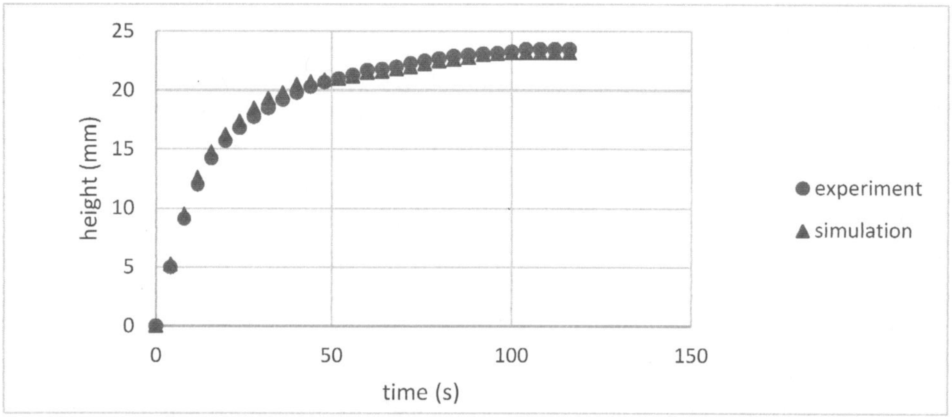

Figure 2. Height of the capillary fluid for different times is plotted for both experiment and simulation.

According to equations (6) and (7)

$$RMSD = \sqrt{\frac{\sum_{t=1}^{n}(x_{1.t}-x_{2.t})^2}{n}} \tag{6}$$

$$NRMSD = \frac{RMSD}{x_{max}-x_{min}} \tag{7}$$

root-mean-square deviation (RMSD) of the results is 0.42 and normalized root-mean-square deviation (NRMSD) is 1.83%, which means that the data obtained from the simulation is very close to the one obtained from the experiment.

3.3. Changing the geometry

To investigate the effect of the tube's shape on the capillary fluid rising, we changed the capillary tube's outer wall to two sinusoidal walls. The shape of walls is described by equation (8) and (9).

$$r = -0.025 * sin(2 * pi * z) + 0.6 \tag{8}$$

$$r = -0.050 * sin(2 * pi * z) + 0.6 \tag{9}$$

Wavelengths of the both sines are the same and the difference is just in their amplitude which affects the diameter of the tube as a function of height. Mean radius of walls and length of the both tubes are 0.6mm and 25mm, respectively. Meshes for both new shapes are depicted in the Figure 3.

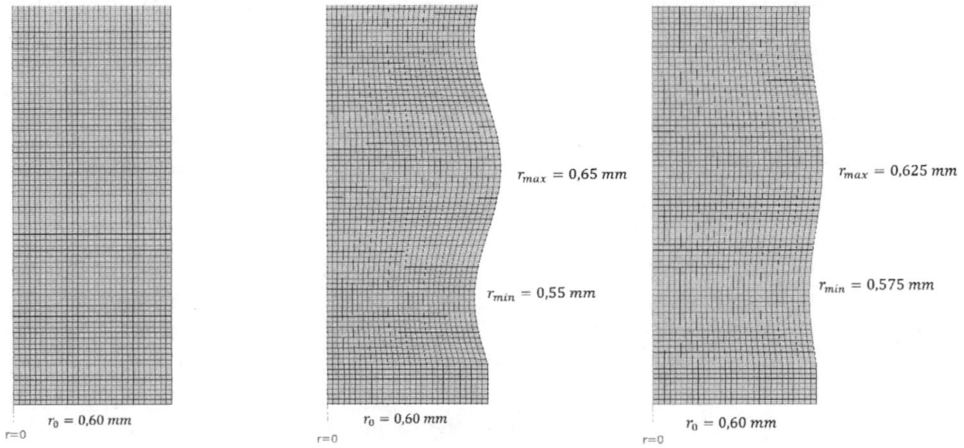

Figure 3. Meshes for different geometries.

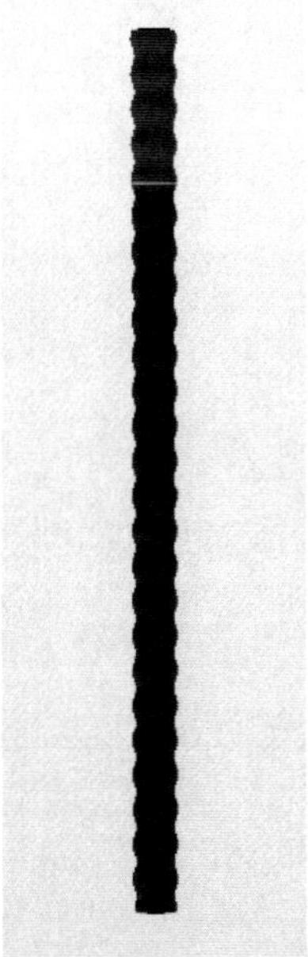

Figure 4. Final result for sinusoidal wall.

Because of the fact that the fluids and the general physics of the problem are the same, there is no need to change the boundary conditions for new shapes.

4. Results and discussion

Computation of the new geometries has been done and an image of the final result for a sinusoidal wall is depicted in Figure 4.
In order to compare the results better, height and time are scaled using equations (10) and (11).

$$Z^* = \frac{z}{h} \tag{10}$$

$$t^* = \frac{t}{\frac{8\mu_c h}{\Delta\rho g r^2}} \tag{11}$$

Where r and h and μ_c are radius of tube, final height and the larger viscosity, respectively [8].

Scaled height of the imbibing fluid versus scaled time for both of the new wall types is plotted in a log-log diagram (Figure 5).

Changing the shape of the wall has affected the capillary rising in different aspects. For simple capillary tube, the nonlinear pattern in capillary rising is obvious. According to *Walls et al*[7], the height of the capillary fluid for the condition which the viscosities are equal, is an exponential function of time (equation (12)).

$$Z = h(1 - e^{\frac{-\Delta\rho g a^2}{8\mu L}t}) \tag{12}$$

But by changing the geometry, the fluid didn't follow the same pattern and the regime of motion changed. The necessary times for the fluid to reach the equilibrium in these new geometries is decreased in comparison with the simple tube, and are so close to each other.

By fitting linear curves on the sinusoidal wall's data, r-squares for fitted lines were 0.99 (for wall with maximum radius of 0.65mm) and 0.98 (for wall with maximum radius of 0.625mm). The slopes of the lines in the log-log plot was 0.89 and 0.93, so the height of the capillary fluid will have a polynomial relation with time (equations (13) and (14)) for these sinusoidal wall capillary tubes.

$$Z_{(r_{out}=0.65mm)} \propto t^{0.89} \tag{13}$$

$$Z_{(r_{out}=0.625mm)} \propto t^{0.93} \tag{14}$$

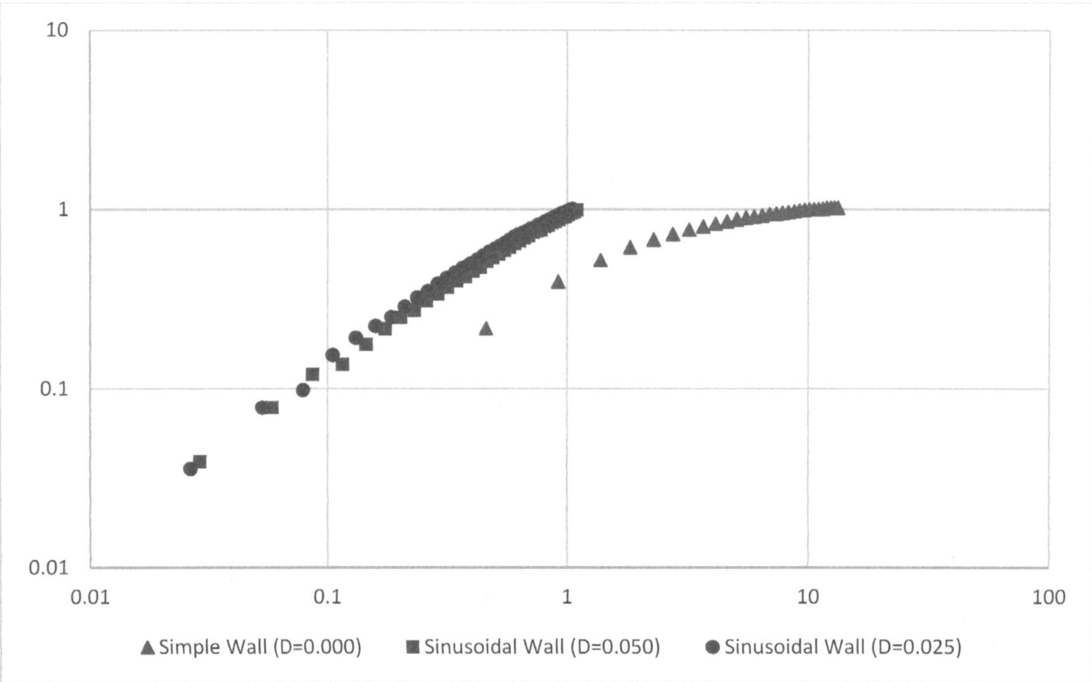

Figure 5. Height of the penetrating fluid versus time for different geometries.

5. Conclusion

Motion of fluid in a capillary tube is a function of many parameters. Viscosity of the fluids, gravity and tube's shape are three important ones. Although many researches have been done about vertical capillary rising, most of them neglected at least one of those three. In this paper, capillary tube's shape effect on fluid rising for a sinusoidal geometry is investigated by considering both gravity and viscosity of the displacing fluid. The results are compared to a simple capillary tube's data. According to them, shape of the capillary tube can affect not only the necessary time for the fluid to reach the steady state, but also can change the fluid motion regime. By changing the geometry, the penetrating fluid does not follow the same pattern and the regime of motion changes. By fitting linear curves on the sinusoidal wall's data, the slopes of the line in the log-log plot are 0.89 and 0.93, so the height of the capillary fluid had a polynomial relation with time which was described by equations 13 and 14.
Nowadays, capillary tubes can be used for variety of applications. Thermal mass flow metering [9], measuring viscosity [10], Microbial detection of enrofloxacin in milk [11] and aerosol generating [12] are some them. Capillary tubes are also used for destruction of volatile organic compounds used in a semiconductor industry [13] and self-assembly of rod like bio-nanoparticles [14]. We understood that new geometries affect the rising regime in a vertical capillary tube, and according to their various applications, this new phenomenon can be used in the future for manufacturing better equipment.
Although many of the parameters that have effect on the fluid capillary motion are studied until now, but it seems that there is a long way to completely understand the behavior of the fluid in this field.

Acknowledgments
The authors thank Sharif University of Technology and Sharif High Performance Computing Center staff for their help and support.

6. References

[1] Hooke, R. An Attempt for the Explanation of the Phenomena Observable in an Experiment Published by the Honourable Robert Boyle. Printed by J.H. for Sam. Thomson, London 1661.
[2] Parvizi, S., A. Arabkoohsar, and M. Farzaneh-Gord. "Natural gas compositions variation effect on capillary tube thermal mass flow meter performance." Flow Measurement and Instrumentation 50 (2016): 229-236.
[3] Kim, Wan Joong, et al. "Portable viscometer and method of manufacturing capillary tube for measuring viscosity." U.S. Patent Application No. 15/013,926.
[4] Lee, Hyeonjeong, et al. "Microbial respiration-based detection of enrofloxacin in milk using capillary-tube indicators." Sensors and Actuators B: Chemical 244 (2017): 559-564.
[5] Cai, Jianchao, et al. "An analytical model for spontaneous imbibition in fractal porous media including gravity." Colloids and Surfaces A: Physicochemical and Engineering Aspects 414 (2012): 228-233.
[6] Cai, Jianchao, et al. "Generalized modeling of spontaneous imbibition based on Hagen–Poiseuille flow in tortuous capillaries with variably shaped apertures." Langmuir 30.18 (2014): 5142-5151.
[7] Walls, Peter LL, Gregoire Dequidt, and James C. Bird. "Capillary displacement of viscous liquids." Langmuir 32.13 (2016): 3186-3190.
[8] Zhmud, B.; Tiberg, F.; Hallstensson, K. Dynamics of Capillary Rise. J. Colloid Interface Sci. 2000, 228, 263–269.
[9] Parvizi, S., A. Arabkoohsar, and M. Farzaneh-Gord. "Natural gas compositions variation effect on capillary tube thermal mass flow meter performance." Flow Measurement and Instrumentation 50 (2016): 229-236.
[10] Kim, Wan Joong, et al. "Portable viscometer and method of manufacturing capillary tube for measuring viscosity." U.S. Patent Application No. 15/013,926.
[11] Lee, Hyeonjeong, et al. "Microbial respiration-based detection of enrofloxacin in milk using capillary-tube indicators." Sensors and Actuators B: Chemical 244 (2017): 559-564.

[12] Price, Shane, et al. "Bent capillary tube aerosol generator." U.S. Patent Application No. 15/434,868.
[13] Kohno, Hitoshi, et al. "Destruction of volatile organic compounds used in a semiconductor industry by a capillary tube discharge reactor." IEEE Transactions on Industry Applications 34.5 (1998): 953-966.
[14] Lin, Yuan, et al. "Self‐Assembly of Rodlike Bio‐nanoparticles in Capillary Tubes." *Angewandte Chemie International Edition* 49.5 (2010): 868-872.

Institute of Physics
Dirac House, Temple Back
Bristol BS1 6BE UK

ISSN: 1757-8981
ISBN 978-1-5108-6281-4

CURRAN ASSOCIATES INC.
proceedings.com